Managing Advanced Manufacturing Technology

The Challenge of the Fifth Wave

John Bessant

NCC Blackwell

MANCHESTER • OXFORD

First published 1991

NCC Blackwell Ltd
108 Cowley Road, Oxford, OX4 1JF, UK

Editorial Office: The National Computing Centre, Oxford House,
Oxford Road, Manchester M1 7ED, UK

British Library Cataloguing in Publication Data
A CIP catalogue record for this book is available from the British Library.

Library of Congress Cataloging in Publication Data
Bessant, John R.
    Managing advanced manufacturing technology : the challenge of the
fifth wave/John Bessant.
        p.   cm.
    ISBN 0–631–16221–6 : $40.00. — ISBN 0–631–16222–4 : $14.95
    1. Computer integrated manufacturing systems.
    2. Just-in-time systems.
    3. Flexible manufacturing systems.
    I. Title.
TS155.6.B483   1991
670.42--dc20
                                                        90–23489
                                                        CIP

Typeset in 11 on 12 pt Times Roman by Bookworm Typesetters, Manchester.
Printed in Great Britain by Hobbs the Printers of Southampton.

. . . For Tom

Contents

v

Preface

This book is targeted at practising managers and students of management, in order to attempt to put into perspective the radical changes now under way in the field of manufacturing technology, and to draw out the challenges which these open up for management in the 1990s and beyond.

This book has two dominant themes. The first seeks to explain the key technologies associated with the new patterns of 'best practice' in manufacturing, and the forces which have led to their emergence. The second is to explore early experience in their application and to examine the kinds of organizational and managerial learning needed to exploit them effectively. Much recent discussion about best practice suggests that successful exploitation of technological opportunity depends not only on the technical system but on the match between that and the social and institutional framework.

The argument goes as follows: As major new technologies diffuse, so what is held as 'best practice' in industrial organization is challenged and disrupted. The old model becomes increasingly inappropriate and counterproductive and experiments take place to try and develop more suitable alternatives. These eventually lead to a new approach, which has been called 'a new techno-economic paradigm'.

This interpretation to helps explain why there should be a shifting geographical focus for economic growth linked to technology. Although technologies may be similar, the social and institutional infrastructures into which they are introduced will vary; different countries will have different (and more or less appropriate) patterns. What is clear – and has been since Schumpeter first commented on it in the 1920s – is that technological innovation cannot proceed successfully without parallel managerial and organizational change. There is a need for an element of 'creative destruction' to enable the evolution of new models for organizing and managing, and this book represents an attempt to track this process.

This book has three strands. The first describes some of the key components of the current revolution in manufacturing technology and the experience gained so far with their use. Thus there are chapters on computer-aided design, on programmable control in production, on computer aids to production management, on new non-computer techniques such as just-in-time manufacturing and total quality control, on new patterns of inter-firm relationship and, finally, on computer-integrated manufacturing. Throughout the book, emphasis is placed upon the more advanced and *integrated* applications of new technology, those able not only to help us do a little better what has always been done, but also to do it in new and often radically different ways.

The second strand attempts to identify the key issues influencing the adoption and implementation of these technologies. This is done through the use of case studies and research material based on our work at the Centre for Business Research and on many other studies carried out by colleagues in the UK and overseas. Each chapter contains a review of such experience and contrasts the *potential* gains from new technology with what has *actually* been achieved. Themes raised in these chapters are drawn together in a concluding section which looks systematically at the managerial and organizational changes needed to exploit the potential of the powerful new technologies of the 1990s.

Finally, the book has a prescriptive strand. Although, to a large extent, each firm is unique, there are some common guidelines for defining strategy, identifying needs and implementing technology, without which investments in advanced manufacturing technology are unlikely to succeed.

JOHN BESSANT

Acknowledgements

This book has evolved out of a number of research projects and teaching activities at Brighton Polytechnic over the past eight years and I would like to take this opportunity to thank the numerous friends and colleagues who have helped to shape and influence my thinking and writing. In particular I would like to extend special thanks to Joanna Buckingham, John Ettlie, Chris Freeman, Bill Haywood, Raphie Kaplinsky, Richard Lamming, Paul Levy, Clive Ley, Carlota Perez, Howie Rush, Paul Simmonds, Stuart Smith, Joe Tidd, David Tranfield and Sally Wyatt, who have, at various stages, read and commented on drafts, or been part of the research which underpins the book. Thanks are also due to the many friends and colleagues in Venezuela, and Cyprus with whom I worked out some of the underlying ideas in a series of training courses and seminars, and to members of the IIASA CIM programme network. My colleagues at the Centre for Business Research and in the Business School have been very patient and encouraging, and particular thanks are due to our superb administration team (Carolyn, Julia, Ouida, Pauline and Marion) who helped in various ways to run things in the department while I was hiding away and writing!

Thanks are also due, for financial support and encouragement, to the Joint Committee of the Economic and Social Research Council and the Science and Engineering Research Council, and to the ACME Directorate of the SERC.

Of course, the biggest debt I owe is to my family, and especially my wife, Norah. They have put up with a great deal while I was locked up in my ivory tower trying to write, and without their patience and support the book simply could not have been completed.

1 Towards Factory 2000

1.1 INTRODUCTION

In 1850 Great Britain was known as 'the workshop of the world'. This was no idle boast: at that time world trade in industrial goods was dominated by Great Britain's production of, for example, 50 per cent of the world's iron, 50 per cent of the world's cotton cloth, 70 per cent of the world's coal and 40 per cent of the world's manufactured goods.

Yet, just over a century later, the UK has a massive £10 billion deficit in manufactured goods and a persistent reputation for low productivity, poor quality and inconsistent and late delivery performance. Whereas in 1900 Great Britain built 60 per cent of the world's shipping fleets, the UK now accounts for less than 3 per cent. In 1948 the UK was the third largest steel producer, but now only just manages tenth place. The machinery industry, once dominated by the UK, is now massively penetrated by imports: only 3 per cent of machine tools and 8 per cent of textile machinery is produced domestically. Over 50 per cent of cars and 98 per cent of motorcycles are imported, together with similar percentages for consumer goods of all types.

Only one other nation amongst major industrial countries – the United States – has a worse record on both post-war productivity growth and current trade deficit. Here too we see the pattern of rise and then fall of industrial fortune, leading to a state of crisis. From being the pre-eminent post-war economy the US has slumped in manufacturing to a position in which there is now a substantial trade deficit in manufactured goods, much of it with Japan. In 1986 imports exceeded exports by $140 billion, and the US role as the world's most powerful exporter passed instead to the Federal Republic of Germany.

By contrast, Japan has consistently demonstrated its strength as a manufacturing nation. From the early 1970s onwards, it has achieved market penetration and dominance in a series of manufacturing industries, despite the growing problem of a high-valued yen. For example, in the early 1980s with an exchange rate of around ¥250 to the dollar, Japanese manufacturers reckoned that they could cope with

1

changes up to a level of ¥180 to the dollar. By 1988 the rate was ¥128 and Japan was still successfully exporting to the US – and estimates suggest that the current (1990) level of competitiveness at which Japanese manufacturers can sell into the US is around ¥150 to the dollar, with some firms claiming even greater competitiveness.[1] This continued success is related to consistently good performance, not only in keeping costs low but also in adding value to products through non-price factors such as quality, design, delivery performance and service.

Comparisons in particular industries highlight this further. For example, in the car industry a recent survey concluded that the average Japanese plant can produce a car of similar complexity and specification with only half the human effort (managerial and shop-floor) required in European plants. This gap has persisted throughout the 1980s and shows signs of widening. Although the position is improving with regard to US car manufacturers, the cost difference is still large. Throughout the early parts of the 1980s it was estimated to be about $2000 per car cheaper to build in Japan than the US.[2] Data from several major industrial nations, presented in figure 1.1, highlights the relative performance gap which is opening up.

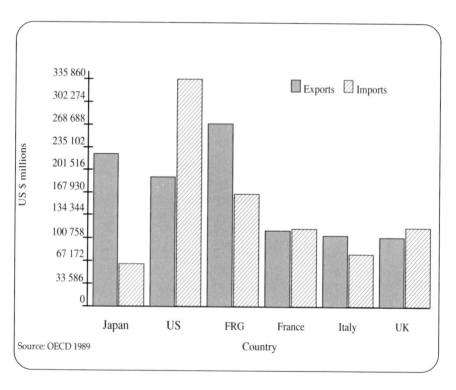

Figure 1.1   Imports and exports of manufactured goods

What has happened to economies such as those of the UK and the US to explain their recent poor performance? And, on the other hand, how have other nations managed to achieve considerable success in post-war manufacturing, with strong productivity, growth, an enviable record on exports and a reputation for high quality and good design? Two such nations in particular – Japan and West Germany – began with marked disadvantages. They ended the 1940s with their economies in tatters and their factories in ruins. The reasons for their success – or indeed, for the failure of the UK and US manufacturing economies – are complex and widely discussed. But one element is of interest to us here: how each of these nations was able to mobilize and use the weapon of technology to greater or lesser effect.

Right back to the first Industrial Revolution, technology has been a critical element in economic growth – a fact which is still acknowledged by economists today. Much of productivity growth is related in some way to 'technical progress', and while our understanding of this is imperfectly developed, the links identified by Schumpeter and Marx are still valid. So the question of how we manage technology and exploit this force for economic growth is critical.

1.2   TYPES OF TECHNOLOGICAL CHANGE

It is important to differentiate technological changes in terms of their novelty and the extent to which they represent a change over what went before. A spectrum of innovation, based on analysis of a large database of twentieth century innovations, is shown in table 1.1.[3]

Type	Characteristics
Incremental	Day-to-day improvements and modifications
Radical	Discontinuous leaps in product or process technology; for example, float glass, nylon, and the compact cassette
New technology system	Far-reaching changes affecting more than one sector or giving birth to new sectors, such as synthetic materials – often accompanied by organizational innovations
New techno-economic paradigm	Major shifts affecting the entire economy, involving technical and organizational change, changing existing products and processes and creates new industries, and establishing the dominant regime for several decades

Table 1.1   Different types of innovation

At one level it is possible to identify many small improvements, each of which represents an increment of change and a gradual substitution process. These take place on a day-to-day basis as firms learn to use and adapt technology to their needs, although their importance in economic growth should not be underestimated. Although each step may be small, their cumulative effect over time can be significant as – for example – shown by Muller in the case of DuPont's rayon plants and Enos in the petroleum refining industry.[4]

Beyond such regular 'bread and butter' changes it is possible to identify innovations which are much more far-reaching in their impact on a product or process. Radical innovations of this kind occur less frequently, but have a correspondingly greater impact when they emerge. They represent a discontinuous stage of development, often opening up new approaches in product or process design. Examples of radical innovation in manufacturing processes include the development of the float glass process or the basic oxygen process for steelmaking, while radical product innovations include the compact cassette, antibiotic drugs or nylon fibre.

The next stage on this continuum involves changes in a technological system, in which a sector or group of industries is transformed by the emergence of a new technological field. Such changes are often accompanied by changes in the organization of production both within and between firms. Examples here include the radical changes brought about by the emergence of new synthetic materials in the post-war period.

Lastly come technological changes which involve not only changes in technology but also in the social and economic fabric in which they are located. Such 'revolutions' do not occur frequently, but their influence is pervasive and long-lasting. For example, the role of steam power as a technology was not confined to radical improvements in the coal-mining industries. It was the catalyst for the Industrial Revolution, and its development and application set patterns which dominated economic growth for many decades to follow.

1.3   THE 'LONG WAVE' MODEL

Such technological revolutions can be associated with other technology clusters such as the development of the railways, the internal combustion engine and electricity. Attention was first drawn to them and their influence on economic growth by a Russian economist, Kondratiev, who argued for the existence of 'long waves' of economic growth and recession. Recent years have seen considerable development and elaboration of these ideas and, in particular, have led to an improved understanding of what actually takes place and how this may be interpreted in terms of 'long waves'.[5]

In this discussion, four such long waves can be identified, each having the following characteristics:

1   They involve a cluster of key technologies which share the characteristics of widespread availability, rapidly falling costs and pervasive potential for application throughout industry and commerce. Taken together, these clusters represent significant growth opportunities in both existing and new industries, and thus fuel economic expansion. The four clusters of technology are represented in figure 1.2.

2   The expansion which they give rise to eventually runs out of steam as the opportunities are taken up, and emphasis turns to cost-reducing applications of the technology and to a downturn in economic growth. Such periods of alternate growth and decline represent the upswing and the downswing of the long wave.

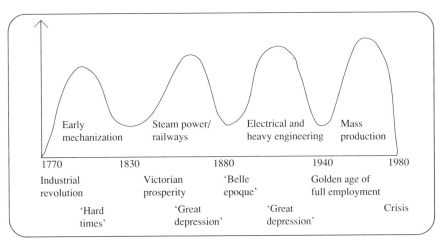

Figure 1.2   Major technological clusters and long waves of economic activity

3   The emergence of economic growth comes primarily through established 'carrier' industries which are able to exploit the opportunities opened up by the technology clusters. In doing so they make available key commodities which fuel economic growth elsewhere in the economy. For example, the second wave revolutionized the coal industry and, through the vehicle of steam power, offered a widely applicable source of power for a variety of applications. Similarly, the period following the Second World War was characterized by the availability of cheap energy, particularly that based on oil.

4   During each cluster new industries also begin to grow and, although only in embryonic form in the current wave, these play a key role as

carrier industries in the next. For example, the development of low-cost steel was only in its infancy during the second wave, but it played a central role in the third wave which relied upon its widespread availability to underpin the development of new and more physically demanding engineering industries. In this way the seeds of the next wave are already growing in the current one.

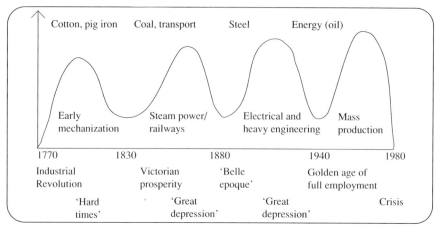

Figure 1.3   Carrier industries and long waves of economic growth

5   Each wave is not only concerned with technological clusters but also has dominant organizational forms associated with it. These set the pattern for organization and management for the next 'era' in manufacturing.

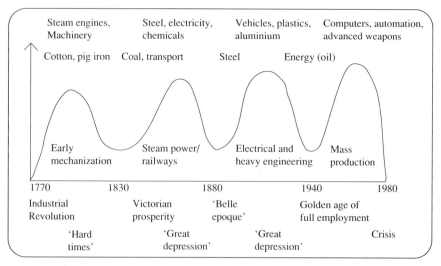

Figure 1.4   Newly emerging industries and long waves of economic growth

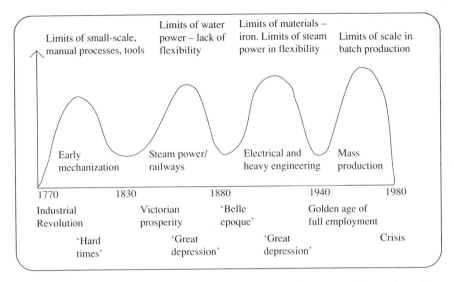

Figure 1.5   Limitations of existing technologies as a trigger for the emergence of a new techno-economic paradigm

6   Finally, the impetus for change does not come solely from the emergence of new technologies (technology push) but also from growing problems and experience of limitations with the

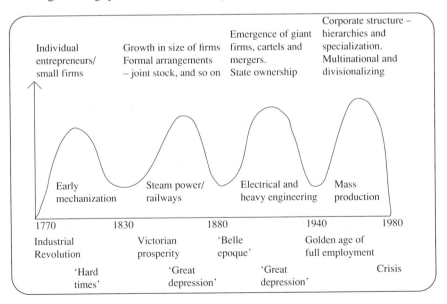

Figure 1.6   Changing organizational and institutional structures and long waves.

technologies and organizational forms associated with the previous wave. Once again, the seeds of the next wave are present in the current one. This also helps to explain the downswing of the wave, where the current pattern of technology and organization beings to prove increasingly inappropriate for dealing with the challenges of the developing economic environment.

This is an oversimplification of the model and a much more detailed account can be found in Freeman and Perez.[6] In particular, they introduce the idea of a 'techno-economic paradigm' in their account of such wave-like changes in the economy. This uses the concept of a paradigm as Kuhn does, to describe a way of seeing, a *'Weltanschauung'*, a dominant pattern which influences thinking across a very broad front. Kuhn has argued that change in science proceeds by a combination of small incremental developments punctuated by periodic 'paradigm shifts' in which the whole structure of how scientists see the world is altered, as for example, in the transition from Newtonian physics to quantum physics. Such a shift brings about enormous opportunities for new thought and sets the dominant pattern for the next period of incremental development.[7]

In an influential paper, Dosi[8] extended the idea of paradigms to the field of technological innovation, arguing that change in this field also proceeds along certain trajectories which are defined by the currently dominant paradigm. Perez[9] took this further using the idea of a 'techno-economic paradigm' which defines the 'commonsense' rules which govern the workings – structural and technical – of industrial society and set the pattern of best practice. These persist for extended periods of time but eventually become increasingly inappropriate; just as, in Kuhn's scientific paradigms, particular ways of seeing become increasingly limited in their ability to explain new observations and experimental results. Eventually the paradigm shifts, a new one emerges, and the cycle repeats itself.

Although there is still considerable debate amongst economists as to the validity of the long wave/paradigm argument, the value of such an approach is that it gives us a model by which to understand some of the changes in industrial economic life, and particularly the role of technological change in the process.

1.4   THE 'FIFTH WAVE': A NEWLY EMERGING PARADIGM?

We can interpret many of the features of late-twentieth-century industrial society in terms of such a shift in paradigms. For example, there is growing agreement that the mass production and mass consumption models which dominated the first half of this century are giving way to more fragmented patterns of demand and more flexible modes of production. Such models are incompatible with older forms of organization, especially those stressing division of labour and rigid bureaucratic organizational forms. Instead we are seeing the emergence of alternative, more flexible arrangements based on networking and decentralization.

The world economy is arguably in a state of transition, in which the previously accepted 'best practice' conditions for industrial performance are changing and – with them – the whole structure of economic society. One indicator of this is the changing pattern of productivity growth, which has declined significantly in Europe and North America but which has been maintained at a high level in other economies, notably that of Japan. There is widespread discussion of a 'crisis' in capitalism, triggered by a variety of forces – the energy crisis of the 1970s, shortages of materials, social unrest, and so on – which is characterized by a growing lack of stability in patterns of investment.

There have been many attempts to explain this change, but all have in common models which suggest a change of 'paradigm', as we discussed above. For example, one group (known as the French regulationist school) suggest that there is in any period of capital accumulation, a balance between the 'regime of accumulation' (which balances consumption, savings and investment) and the 'mode of regulation' (the social and institutional structures within which this takes place). They argue that in the 1970s changes in both of these led to an increasing mismatch, with the result that a crisis for capitalism gradually emerged. In particular, they take the view that the changes in the basic conditions of accumulation – with fragmenting markets and shifting demand patterns – became increasingly out of line with the rigid organizational structures and practices associated with mass production.[10]

This is a theme also taken up by Piore and Sabel,[11] who argue that as patterns of demand become increasingly fragmented and emphasis shifts to non-price factors, so the necessary industrial flexibility can no longer be offered by mass production systems:

Mass production offered those industries in which it was developed and applied enormous gains in productivity – gains that increased in step with the growth of these industries. Progress along these technological trajectories brought higher profits, higher wages, lower consumer prices and a whole range of new products. But these gains had a price. Mass production required large investments in highly specialised equipment and narrowly trained workers. In the language of manufacturing, these resources were 'dedicated'; suited to the manufacture of a particular product – often, in fact, to make just one model. Mass production was therefore profitable only with markets that were large enough to absorb an enormous output of a single, standardised commodity, and stable enough to keep the resources involved in the production of the commodity continuously employed.

They suggest an alternative model, that of 'flexible specialization', which emphasizes the production of smaller batches of higher variety, through new flexible technology and alternative organizational arrangements.

In parallel with these developments the newly emerging industries of computers and the industrial automation of the 1950s have now matured to the point at which information technology (IT) offers a pervasive, low-cost resource which has extremely wide potential applicability. Close behind IT are both biotechnology and new materials technologies, while in the longer term the emerging industries of the twenty-first century may well be associated with further developments in space exploration (including new manufacturing techniques).

Therefore the scene is set for a shift in the dominant techno-economic paradigm, one which involves not only technological change but also institutional change, pulled through by failures of the existing paradigm as much as pushed by new technological opportunities.

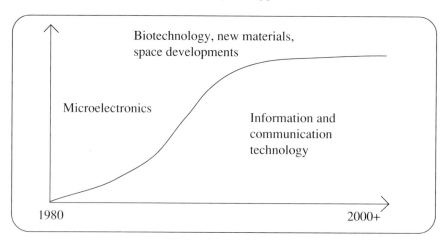

Figure 1.7   The fifth wave?

1.5   THE TRANSITION TO POST-FORDISM

There is growing acceptance of some version of this paradigm shift model amongst economists and policy-makers. For convenience, the label 'Fordist' has been used to describe the old and increasingly inappropriate model based on mass production and rigid, bureaucratic organizational forms. The contrasting 'post-Fordist' paradigm is by no means clearly defined, but includes a number of features such as:

- greater emphasis on non-price factors
- greater emphasis on flexiblity in technology
- greater emphasis on flexibility in organizational structure
- changing relationships within and between organizations

Perhaps the key question for manufacturing enterprises as they move into the 1990s is how to effect the transition. Two elements would appear to be crucial for success: change based on the new technologies of information and communication, and organizational change towards new supporting structures. (There already seems to be some support for this view. One explanation for Japan's strong economic performance may lie in the fact that they have developed and used alternative organizational forms and have been early and extensive users of new information-based technologies.)

The problem facing firms wishing to change is that there is no clear blueprint available to define the new organizational forms or technological configurations that are needed – any more than earlier paradigms were clearly articulated at their start. Instead, there is a need for experimentation, innovation and learning, as new options are tried and evaluated as models for 'best practice' as alternatives to the increasingly inappropriate 'Fordist' model. One of the phrases beginning to enter the manufacturing language of the 1990s is the concept of 'the learning organization'. Given the challenge of trying to find and develop a new paradigm, this ability to learn may become the critical skill that will determine competitiveness in the future.

Notes

1   N. Valery, 'Can Japan still make it?', *The World in 1990*, (Economist Publications, London, 1990).
2   D Jones and J. Womack, 'The real challenge facing the European motor industry', *Financial Times*, 28 October 1988.
3   Based on studies carried out at the Science Policy Research Unit, Sussex University and on analysis of the large innovation database held there.
4   See, for example, J. Enos, 'Measure of the rate of technological progress in the petroleum refining industry', *Journal of Industrial Economics*, 6, 1958, pp.180–97, and S. Hollander, *The Sources of Increased Efficiency: A Case Study of du Pont Rayon Manufacturing Plants*, (MIT Press, Cambridge, 1965).
5   For a detailed discussion, see C. Freeman, J. Clark and L. Soete *Unemployment and Technical Innovation; a Study of Long Waves in Economic Development*, (Frances Pinter, London, 1983).
6   C. Freeman and C. Perez, 'Structural crises of adjustment, business cycles and investment behaviour', in *Technical Change and Economic Theory*, ed. G. Dosi (Frances Pinter, London, 1989).

7   T. Kuhn, *The Structure of Scientific Revolutions* (University of Chicago Press, Chicago, 1962).

8   G. Dosi, 'Technological paradigms and technological trajectories', *Research Policy*, 11, (3) (1982), pp.147–62.

9   C. Perez, 'Structural change and the assimilation of new technologies in the economic and social system', *Futures*, 15, (4) (1983), pp.357–75.

10 For a discussion of the 'French regulationist school', see M. Aglietta, *The Theory of Capitalist Regulation*, New Left Books, London, 1976.

11 M. Piore and C. Sabel, *The Second Industrial Divide: Possibilities for Prosperity*, (Basic Books, New York, 1984).

2 The New Manufacturing Challenge

2.1 THE NATURE OF NEW TECHNOLOGIES

New technologies often seem to offer powerful solutions to complex problems, but this leads us to think of them as being packed in a bright new box on the supermarket shelf, waiting for us to collect and use them. Unfortunately, such technologies do not simply appear magically, from out of nowhere. They are the product of a complex web of interacting forces, pushing and pulling, shaping their direction and the speed and form in which they emerge. They are simultaneously the product of a number of needs drawing a technological response *and* the key to new opportunities, new possibilities which open up as a consequence of their emergence.

A new technology is also not a single product destined for a single customer: rather, it is a *field of opportunity*, offering a wide range of choices to suit an equally broad range of users. There is no guarantee that the right technology will be chosen, or even that a particular firm will find out about its existence. Researchers have explored these ideas in some depth, and have highlighted the ideas of 'technological trajectories' (the complex way in which particular technologies emerge at a particular time and in a particular form) and 'search environments', the space (which varies between firms) in which potential users explore for technological opportunities.[1]

Others point out that the process is not a sequential one in which technology evolves and user organizations adapt, but rather an interactive one in which technologies shape the way organizations use them which in turn shapes the next stage of development. Organizational and management characteristics may not simply inhibit or facilitate the diffusion and implementation of a technology; they may also play a key role in its development and evolution. Such a 'configurational' model is particularly applicable to the current generation of complex and highly integrated manufacturing technologies which represent *systems* rather than single elements of technology.[2]

Today's advanced manufacturing technologies offer enormous potential benefits in all aspects of factory operation and beyond. But unless we are aware of the complex way in which they have evolved and become available, and of the needs to which they are a response, it is unlikely that we will be able to manage them effectively. Simply throwing a technology at a problem because it happens to be available is a high-risk gamble which often fails to pay off. In the end a technology is simply a tool (albeit often very powerful), and in order to make the best use of that tool we need to understand it – not treat it like a magic potion to be applied randomly to treat industrial ills.

There are two components to this understanding. The first is to appreciate the reasons behind the emergence of a particular technology. What forces have shaped it, pushing and/or pulling it into the form with which we are now familiar? This will help to identify the kinds of problem to which it can be appropriately applied, and its limitations within that context.

The second is to understand that its development is not fixed once, for ever, but is continuous. Users often treat technology as something over which they have no control beyond the decision to purchase and install it. Yet the role of the user is anything but passive in practice; the experience of using and adapting a technology is a key feature that shapes the rate and direction of its future development.

2.2 TECHNOLOGY AS A STRATEGIC WEAPON

We often treat technology simply as one more tool for a particular task, able to help us do that one job a little better. This limited view fails to recognize the true potential of technology as a *strategic* resource, something which can help a firm in pursuit of its long-term strategic goals. In this respect technology represents a powerful weapon – but one which can only be effective if it is pointed in the right direction and used by someone who understands its capability and how best to exploit it.

Business strategy (see figure 2.1) is essentially formulated in response to a careful analysis of the business environment; its threats and opportunities and the relative strengths and weaknesses of the firm. Such an analysis needs to take account of the many different actors and interest groups – customers, competitors, government, pressure groups, and so on – and of the dynamics whereby the relationships between these can change rapidly.[3]

It involves a number of interrelated key areas, including finance (how the business will be funded and controlled), marketing (how the products/services will be advertised, distributed and sold) and product (what the business will make for which markets) (see figure 2.2).

Recent years have seen a growing emphasis on manufacturing strategy as an important component of competitive behaviour, as attention has shifted from simply selling on price to looking at other characteristics, such as quality or delivery, as well. Hill makes the important distinction between 'order qualifying criteria', which are the things a firm must be able to offer in order to even be considered by a potential customer, and 'order winning criteria', which are those characteristics which beat the competition. A definition of manufacturing strategy is thus the ways in which order qualifying and order winning can be achieved.[4]

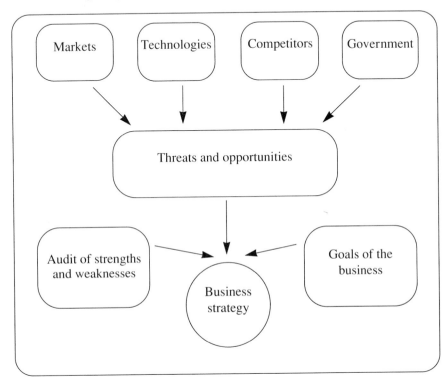

Figure 2.1   Components shaping business strategy

Manufacturing strategy is primarily concerned with how the products will be made, or the services delivered, and includes factors such as:

- choice of process or technique
- make or buy decisions
- setting quality standards and procedures
- production and work organization
- requirements for physical facilities (buildings and services)
- design of planning and control systems
- investment plans and justification

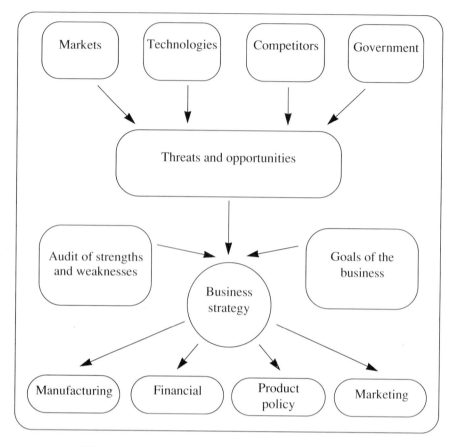

Figure 2.2   Components of a business strategy

One of the most significant tools in manufacturing strategy is technology, the combination of equipment, software and organization which facilitates manufacturing. And there is a direct connection whereby what happens in the environment is critical to shaping technological development – in pulling it in particular directions – while technology in turn is critical in defining new opportunities in the market environment. An understanding of the nature and dynamics of this relationship is central to effective management of technology.

2.3   LOOKING BACK. . .

We can begin to develop this understanding if we begin to track some of the historical changes in technology and the environments which spawned them. At the turn of the century manufacturing saw a period of dramatic expansion. New industries were blossoming as entrepreneurs sought to provide people with a range of new products and services never before available on a large scale – cars, home entertainment, new food products, and so on. As the world moved through the increasing militarization which culminated in the First World War the expansion in military products was even more dramatic, including new machines, armour plating, aircraft, chemical weapons, explosives, food preservation, large-scale manufacture of clothing and various synthetic substitutes for scarce items. When the war was finally over, the mass production capacity for these industries had to seek other outlets, providing another stimulus for market expansion and development.

For a manufacturer supplying such rapidly growing markets, the key concern was output – to produce as much as possible at as keen and competitive a price as possible. Emphasis was placed on economies of scale, the theory being that the bigger the plant, the cheaper it became to produce a single unit of product. Typical of the products available at this time – and in many ways a symbol of the whole era – was Henry Ford's Model T. A car for 'Everyman', it represented the triumph of the manufacturing system over small-scale craft production; and its manufacture required a new pattern of production technology and organization.

The origins of this new pattern can be traced back to Adam Smith's famous observations of the pin-making process, which marked the emergence of the concept of the division of labour. By breaking up the task into smaller, specialized tasks performed by a skilled worker or a special machine, productivity could be maximized. During the next hundred years or so considerable emphasis was placed on trying to extend this further, by splitting tasks up and then mechanizing the resulting smaller tasks wherever possible to eliminate variation and enhance overall managerial control. The ideas of industrial theorists such as Charles Babbage and Andrew Ure, both of whom saw the factory as a complex machine in which people were to be seen as parts, inter-changeable with and replaceable by machines, were influential here.

The industrial engineer, Frederick Taylor, with his *The Principles of Scientific Management*[5] extended this principle, stressing the need to remove discretion wherever possible: 'all possible brain work should be removed from the shop and centred in the planning and laying out department.' His principle of layers of functional management introduced the separation of indirect as opposed to direct workers and the pyramidical structure of management.

It is easy to underestimate the importance of this new approach to manufacturing. In many industries (such as metalworking and engineering) it represented a radical departure from previous practice in which highly skilled and trained craftsmen produced goods in small batches. Although the Industrial Revolution of the previous century had brought the 'factory system' into being and had moved manufacturing some way from being a craft activity towards being organised and machine-based, there was still much in the organization and use of resources which was inappropriate to the demands of high-volume production.

Whereas the British tradition had largely been one of 'making', it was US industry which really laid down the basic ideas behind 'manufacturing'. The ideas of Eli Whitney, originally introduced to carry out high-volume gun manufacturing in the early 1800s, made an important contribution here and gave birth to factories such as that operated by Samuel Colt in Hartford in which:

> . . .the whole floor-space is covered with machine tools. Each portion of the firearm has its particular section. As we enter the door the first group of machines appears to be exclusively employed in chambering cylinders; the next is turning and shaping them; here another is boring barrels; another group is milling the lock-frames; still another is drilling them; beyond are a score of machines boring and screw cutting the nipples, and next to them a number of others are making screws; here are rifling machines, and there the machines for boring rifle barrels . . . nearly 400 are in use in the several departments.[6]

In many ways, this was the forerunner of the functionally laid out factory with which we are familiar today.

This 'American system' stressed the notion of the 'mechanization of work'. As Jaikumar puts it:

> Whereas the English system saw in work the combination of skill in machinists and versatility in machines, the American system introduced to mechanisms the modern scientific principles of reductionism and reproduceability. It examined the processes involved in the manufacture of a product, broke them up into sequences of simple operations, and mechanised the simple operations by constraining the motions of a cutting tool with jigs and fixtures. Verification of performance through the use of simple gauges insured reproduceability. Each operation could now be studied and optimised.[7]

It was not, therefore, entirely surprising that these same principles were also applied to labour, which was the driving force behind the development of Taylor's ideas of 'scientific management'.[8]

That the convergent blueprint for manufacturing has come to be known as 'Fordism' reflects the enormous influence of Henry Ford in the way in which he (and his gifted team of engineers) developed and systematized such approaches.[9] His model for the manufacture of cars was based on a number of innovations which reduced the need for skilled labour, mechanized much of the assembly process, integrated preparation and manufacturing operations for both components and finished product and systematized the entire process. As Tidd[10] points out, the basic elements of the Ford system were largely already in existence; the key was in *synthesizing* them into a new system. Even the idea of flow production lines for motor cars was first used in the Olds Motor Works in 1902, while Leland's Cadillac design of 1906 won an award for the innovation of using interchangeable, standardized parts.[11] The challenge of the high-volume, low-cost production of the Model T led Ford engineers to extend the application of these ideas to new extremes; involving considerable investment in highly specialized machine tools and handling systems, and extending the division and separation of labour to the point at which workers' main tasks were to feed the machines. The dramatic impact of this pattern on productivity can be seen in the example of the first assembly line, installed in 1913 for flywheel production, where the assembly time fell from 20 to 5 man–minutes. By 1914 three lines were being used in the chassis department to reduce assembly time from around 12 hours to less than 2 hours.

This approach extended beyond the actual assembly operations to embrace raw material supply (such as steelmaking) and transport and distribution. At its height a factory operating on this principle was able to turn out high volumes (8000 cars per day) with short lead times: for example, as a consequence of the smooth flow which could be achieved it took only 81 hours to produce a finished car from raw iron ore – and this included 48 hours for the raw materials to be transported from the mine to the factory![12] In the heyday of the integrated plants such as at River Rouge, productivity, quality, inventory and other measures of manufacturing performance were at levels which would still be the envy even of the best organized Japanese plants today.

Some of the key features of this blueprint for manufacturing, based on Taylor's ideas and typified in the car plants of Henry Ford, but applied to many other industries throughout the 1930s and beyond, are highlighted below:

1   Standardization of products and components, of manufacturing process equipment, of tasks in the manufacturing process, and of control over the process.

2   Time and work study, to identify the optimum conditions for carrying out a particular operation and job analysis, to break up the task into small, highly controllable and reproduceable steps.

3   Specialization of functions and tasks within all areas of operation. Once job analysis and work study information was available, it became possible to decide which activities were central to a particular task and train an operator to perform those smoothly

and efficiently. Those activities which detracted from this smooth performance were separated out and became, in turn, the task of another worker. So, for example, in a machine shop the activities of obtaining materials and tools, or maintenance of machines, or of progressing the part to the next stage in manufacture, or quality control and inspection were all outside the core task of actually operating the machine to cut metal. Thus there was considerable narrowing and routinization of individual tasks and an extension of the division of labour. One other consequence was that training for such narrow tasks became simple and reproduceable and thus new workers could quickly be brought on stream and slotted into new areas as and when needed.

4   Uniform output rates and systematization of the entire manufacturing process. The best example of this is probably the assembly line for motor cars, where the speed of the line determined all activity.

5   Payment and incentive schemes based on results (output, productivity, and so on).

6   The elimination of worker discretion and passing of control to specialists.

7   Concentration of the control of work into the hands of management within a bureaucratic hierarchy, with extensive reliance on rules and procedures – doing things by the book.

2.4   COPING WITH CHANGE

Arguably, Ford's plants represented the most efficient response to the market environment of its time. But that environment changed rapidly during the 1920s, so that what had begun as a winning formula for manufacturing began gradually to represent a major obstacle to change. Production of the Model T began in 1909 and for 15 years or so it was the market leader. As Abernathy points out, despite falling margins the company managed to exploit its blueprint for factory technology and organization to ensure continuing profits. But by the mid-1920s growing competition (particularly from General Motors, with its strategy of product differentiation) was shifting away from trying to offer the customer low-cost personal transportation and towards other design features – such as the closed body – and Ford was increasingly forced to add features. It eventually became clear that a new model was needed, and production of the Model T ceased in 1927.

The change over to the new Model A was a massive undertaking which involved crippling investments of time and money, since the blueprint for the highly integrated and productive Ford factories was only designed to make one model well. The process took a year, during which Ford lost $200 million and was forced to lay off thousands of workers – 60 000 in Detroit alone. Around 15 000 machine tools were scrapped and a further 25 000 had to be rebuilt – and even though the Model A eventually became competitive, Ford lost its market leadership to General Motors.[13]

This highlights one of the key lessons about the manufacturing environment – nothing stays the same for ever! Just as the environment of the 1900s differed radically from that of the nineteenth century, so the emerging picture as we approach the twenty-first century poses new and unexpected challenges. We have seen the massive increase in competition, with many more providers of goods and services, both in the domestic market and overseas. And 'overseas' no longer means simply a handful of firms in the advanced industrialized nations, serving largely dependent and captive colonial/imperial markets. There are now many newly industrializing countries (NICs), competing aggressively for a share not only of their local markets but also in the heartland of the old established manufacturing nations. Japan's model of export-led economic growth provided an example that has been successfully followed by nations on the Pacific Rim (such as South Korea) to the extent that their growth rates are amongst the fastest in the world.

This international position is further complicated by the presence of the massive transnational corporations (TNCs), some of whom have a sales turnover exceeding the GNP of many substantially sized countries. Their elaborate network of manufacturing, sales and distribution operations is spread throughout the world, and their facility to move between geographical locations to secure optimum trading conditions adds another key element to the complex environment of the 1990s.

For the customer, the range of choice of products and services is bewildering, but it is becoming clear that such competition puts a degree of power into to customer's hands. He can now begin to demand better levels of service, better quality products, better delivery and support and greater specificity in what he buys. Such power is often increasingly concentrated in the hands of key groups, such as food retailers, who can exert an enormous influence on manufacturers in terms of what they are asked to produce – and *when* they produce it. Increasing interest has been shown in supply and distribution chain management, with the emphasis on cutting out buffer stocks and inventory holding 'just in case' of problems; instead the move is now towards producing and delivering 'just in time' for something to be used or sold. Moves to create larger markets – such as the 1992 proposals for the creation of a single European market within the EEC – are likely to exacerbate this trend.

Even the products themselves are no longer to be taken for granted. Whereas the life of a typical product might once have been measured in years or even decades, many products life cycles are now down to months. Some consumer product – such as television sets or hi-fi systems – are updated several times every year. It has been estimated that up to 80 per cent of the new products which we will be buying in ten year's time have yet to be invented! The effect of this is to challenge the idea that industries go through phases, moving from being new young and innovative sectors associated with new products to mature industries in which the product and the way in which it is made are well established. Pressure for shorter product life cycles means that industries now need to find means of constant renewal.

All of this poses enormous challenges for manufacturers – and takes them a long way from the world in which Henry Ford was operating 80 years ago. Perhaps the key feature is the shift away from an emphasis on price factors and output and towards *non-price* factors. If the world is seen as a market square, then what determines whether a shopper buys from stall A or stall B – assuming that they offer the same price – will be other features such as the quality of the product, the design and packaging, the selling ability of the stallholder, the degree of service which the customer is offered (both during purchase and after sales), and the degree to which the stallholder is able to offer the customer something specifically tailored to their needs rather than simply taken off the shelf as a standard item.

In the UK, for example, the country which was once the 'workshop of the world' moved into a deficit in its trade balance on manufactured goods during the early 1980s. The OECD suggested that, amongst other factors:

> . . . it also seems likely that non-price factors such as delivery times, design and quality worked increasingly against British goods in world markets.[14]

Therefore, keeping the manufacturing customer satisfied moves from being a function of price to one in which such non-price factors are critical, as follows:

- variety – 'any colour you want (as long as it's black!)'
- design
- frequency of product innovation
- quality
- service – before, during and after sales
- delivery
- responsiveness – 'what I want, when I want it'

However, this should not blind us to the traditional concerns within the factory for productivity, the efficient use of inputs of capital, labour, raw materials, energy, and so on to production.

With rising costs of many raw materials, the question of efficient use of these inputs becomes critical. Such questions do not simply relate to using less but also to reducing scrap levels, shortening the production lead time (and hence the time material is held in unfinished form) so that products can be sold faster and money can come back into the business. (To put this in perspective, the cost of holding material in the UK, whether as raw materials, work in progress or finished goods stock awaiting sale, is estimated at between £23billion and £41billion.)

The energy crisis of the mid-1970s helped to focus attention on the problem of energy conservation and efficiency, but while enormous strides have been made, there are still problems for many manufacturers with energy-intensive processes, even during times of relatively low fuel costs.

While the overall proportion of labour involved in manufacturing has declined to 10–15 per cent of the total workforce in most developed countries, the actual costs of employing them have not changed so much.

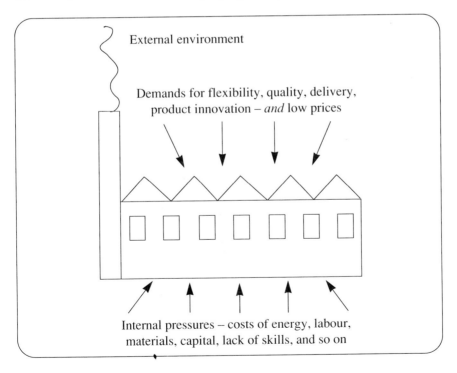

Figure 2.3   Putting the squeeze on manufacturing

Certainly there is less direct labour – and in many industries this is below 5 per cent – but this is often compensated by more indirect, support labour. Additionally, an increasing proportion of wage costs relate to a labour force which is more highly qualified and thus more expensive.

This is the nature of the manufacturing environment in the 1990s, with massive competitive pressure forcing firms to be more flexible, agile, service-oriented and quality conscious – and all underpinned by low price (see figure 2.3). Delivering all this will require a shift in factory organization and technology at least as dramatic as that which took place at the turn of the century.

2.5   THE CHALLENGE

It can be argued that these pressures for change are being felt right across the industrial spectrum, so before we examine the kinds of technological change which are taking place in manufacturing, it will be useful to look a little more closely at how different kinds of manufacturers meet these problems in their particular product or market environment.

Process industries

In the process industries, those industries which might be regarded as the bastion of high-volume, low-variety production, the emphasis has always been on *productivity*; that is, producing as efficiently as possible. Since the product in question has involved little or no variety, this could best be achieved through large plants, in which economies of scale could be obtained.

Economies of scale are based on the principle that the larger the plant (in terms of capacity) the more efficiently a product can be produced. Traditionally in such industries the rule of 1 : 6 has applied: a doubling of capacity has led to a six-fold fall in costs. This benefit has been derived from more efficient use of equipment, of inputs such as energy and labour and in general experience – the well known 'experience curve' effect – which led to the construction of the giant petrochemical and commodity processing plants of the 1960s and 1970s.

Of course, quality and service were important considerations but in essence such businesses were all about high-volume, low-variety production. With such high volumes it became possible to invest in highly specialized technology as well as large plants. As a result many of the key innovations (such as the use of computer control of manufacturing) began life in these industries, the first mainframe computers being installed in sectors such as steelmaking and petrochemicals.

The result of all this investment was that the process industries – typified by the chemicals sector – achieved a consistent and enviable record of productivity growth throughout the post-war period. However, achieving such economies of scale depends, above all, on stability in the marketplace. For as long as demand is increasing (or at least remaining relatively predictable), it is possible to match capacity to this level and ensure efficient production. Similarly, the persistence of demand for a single product or a small range means minimal costs and delays associated with setting up the plant for the next production run. But when the market starts to demand more variety and when volumes for any single product fall (although the total volume may remain steady or even increase) then the economy of scale approach begins to break down.

One consequence is massive over-capacity for many basic items as the huge plants built during the 1970s fail to find sufficient demand, while there has also been considerable growth in small-scale, specialist plant. A good example here is in steelmaking, where the 1970s saw a massive shakeout and rationalization of the industry – a painful process in which thousands of jobs were axed and major capacity reductions took place. The industry is now facing renewed demand, but this time it is no longer concentrated in giant integrated plants but also includes mini-mills and production plants specializing in high-quality, specialist steels and alloys. For example, the share of US production held by the large integrated steelworks fell during the decade 1979–88 from 73 per cent to less than 50 per cent, while the share held by mini-mills doubled, from 10 per cent to 20 per cent.[15] Predictions are that this trend will continue, with a further decline in 'big steel'.[16]

Even where the basic processing can be carried out in large scale plant (such as in sugar-refining) the packaging and finishing operations require major expansion and development to cope with the increased variety and flexibility required. A good example here is that of margarine, a simple product which grew out of wartime shortages of butter and which has since become a major commodity, typical of many in the food processing industry. In the margarine business edible oils are hydrogenated and emulsified, relatively simple processes which used to be carried out in large-scale plants. But now the market demand has fragmented into various different sections – for low-fat products, low-cholesterol products, different oil inputs such as sunflower seed, for the newer high-water products which are blended with buttermilk, and so on. In the UK, Unilever's Van den Berghs division, responsible for margarine and related products currently make some 200 different varieties, and they expect batch sizes to fall further and product variety to increase.

Another example comes from the paper industry; again, traditionally a commodity business based on economy of scale. As Ranta points out, a variety of changes, including energy and material cost increases and the growing concern with environmental issues, have increased pressure on this business. Competition forced a degree of product diversification on the industry in the early 1970s, and this trend has accelerated in recent years with considerable variety now on offer, especially in areas such as tissue and fine papers. Critical competitive advantages depend on non-price factors such as quality, short delivery lead times, customized quantities and fulfilment of special needs.[17]

One final example serves to press the point. In the bulk chemical business, once a commodity operation based on economy of scale, the same patterns can be detected. For example, the supply of carbon disulphide to the textile industry used to be a high-volume, low-variety business; but, as a recent article points out:

> . . . while a decade ago many companies churned out large quantities of commodity materials generally confident they would find customers for them, increased competition and more fragmented markets have in the past few years forced the industry to steer towards more specialised product sectors in which the focus is on tailoring goods to customers needs.

As the managing director of one bulk chemical company commented, this move has necessitated:

> . . . a complete change of philosophy . . . now we have to be far more marketing-led and offer more of a technical service to customers . . . now we have cycle times of a few days and rapid changeover between making different types of materials, which can sometimes lead to technical problems. What we are doing now is really scaled up laboratory work with batches of chemicals going out of the factory in small drums. It's a long way from the days when everything we made at the plant was produced in a continuous stream and went out in tankers.[18]

Similar patterns can be found in the oil industry, where refineries and blending plants are increasingly having to offer smaller batches of customer-specific output. For example, the Shell Lubricants plant at Stanlow, UK, offers a full range of lubricants, in sizes ranging from bottles of motor car oil to barrels of aviation lubricants. The total range involves some 2500 combinations, made up from over 700 recipes blended and then packed into different packages.

What does this mean for the manufacturer? First, the basic production processes need to be scaled down since demand for any single product has fallen. But, alongside this, new processes are needed (such as high-power emulsification to effect the high water blending in our example of margarine). And the increase in variety means that there is a time penalty associated with changing over (cleaning tanks and mixing vessels, and so on) even if the same equipment can be used for different products. This is not only a problem in production processes; it also increases the difficulty of keeping track of everything – of inventory management, production control, and so on – and the risks of production delays which inevitably lead to poor delivery performance.

Mathematically, we can see that even a small increase in product variety creates a much higher level of complexity which has to be managed within the factory, even if the same equipment can be used to make the different products. As new process options are added to this picture, the complexity and uncertainty increase dramatically. The same changes are taking place right across the process industry sector, from petrochemicals through soft drinks to cosmetics. Although the basic production is still high-volume flow processing, the trend is moving fast towards one in which firms have to become more agile and responsive if they are to survive.

Mass production

Manufacturers of some products made in high volumes – such as consumer electronics or motor vehicles – have always had to face the problem of trading off volume against variety. Their production is essentially a batch operation, interrupted by the need to move from different process stages in a discontinuous way. Unlike oil refining, the sub-assemblies and components in a motor car do not flow together and react in a continuous flow manner. Instead, each stage involves different types of operations, different equipment and different components and materials. Since these stages take varying amounts of time, production needs to be planned and each stage carefully integrated into the next, so that the whole simulates smooth and continuous flow as closely as possible. This has led, over time, to the concept of the production line, of techniques such as line balancing and to the development of special-purpose equipment to assist in providing a semblance of continuity and smooth flow.

For example, the transfer line in the car industry brings together different machine tools and kinds of handling equipment so that several operations can be linked togther into a smooth-flowing whole. But such equipment has two drawbacks: it is expensive to install, and if it breaks down the whole line stops. Worse, if the product design changes, the equipment usually needs to be scrapped and a new production line installed, adding further to the cost. Therefore such automation is found only in those industries where volumes are high enough to recover these high capital costs.

But now the same pattern that we saw in the case of the process industries is putting pressure on the mass producers. In the case of motor cars, for example, customers will no longer accept 'any colour as long as it's black'. Indeed, when buying cars they want all sorts of variety; high-performance sports models, the powerful estate car, the luxury executive version, the standard family saloon, the fleet car for companies and car hire firms, and so on. Further evidence of the growing fragmentation of the industry in terms of product offerings is the decision to develop local design centres which will tailor products to particular geographical market preferences.

Such a market environment is unforgiving: any failure to offer models to match those offered by the competition can result in a costly loss of market share. But to provide such variety in the context of fixed production lines and special-purpose equipment geared up to high-volume production of the same good is not easy. This problem becomes more marked when there is no possibility of increasing prices to pass on this cost – indeed, in many cases the pressure on prices is downward. Customers want more quality, variety, and so on, and they want it at a competitive price.

Some indication of this demand for greater variety and shorter product life cycles can be seen in figures 2.4 and 2.5, which highlight increasing flexibility requirements in the Japanese automobile industry.[19]

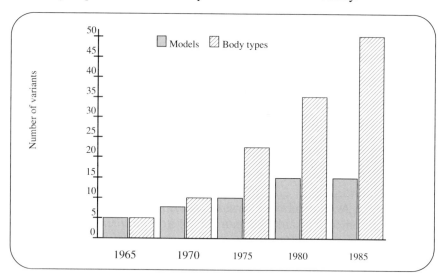

Figure 2.4   Increasing variety in the Japanese motor industry, 1965–85

At the same time, the replacement lives of products are shortening, so that the frequency with which new models are introduced increases. Each time the model changes it potentially requires a new production line with all the associated investment, not only in special-purpose equipment but also in handling and transport equipment, storage facilities, plant layout and – most important – skills development via training. Whereas only a decade ago the life cycle of a car model was around 10 years, the time for new model development is being cut further and further back. In Western firms it is between five and seven years, while in Japan it is around four years, with Honda claiming to have reduced it to three. Estimates suggest that Japanese car manufacturers have been introducing model changes at the rate of 100 per year. [20]

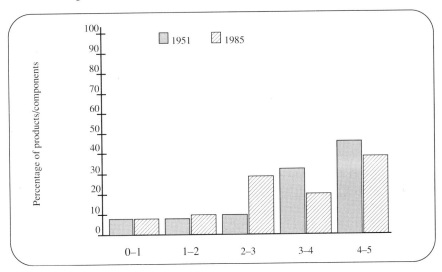

Figure 2.5   Shortening product life in the Japanese motor industry

Furthermore, the break-even size for car manufacturing plants is falling, with Japanese manufacturers now claiming to be able to operate economically producing less than 500 000 cars each year over a four-year period, as compared to Western plants which still need longer (around 10 years) and operate at higher volumes (around two million cars per year).[21]

For industries such as consumer electronics the frequency of changeover is very high, running into several times per year: and while these industries often do not have the same levels of investment in special-purpose equipment, they still have significant problems in changing over assembly, test and other procedures. For example, Milne reports that the typical life cycle of a TV set or video recorder in 1988 was between 10 and 20 months, and that a typical manufacturer would offer between 30 and 40 model varieties, with some offering as many as 70 different products, all targetted at different market segments. [22] Two examples of electronics companies confronting a typical problem of flexibility management are given in table 2.1.[23]

Dimension	Company A	Company B
Total board designs	1400	900
Total boards per year	90 000	40 000
Smallest batch	1	1
Largest batch	100	300
Average batch	30	15
Components per year	14 million	55 million

Table 2.1 Increasing variety in electronics manufacturing

Similar patterns can be found in other consumer-related mass production industries. In the footwear industry, the challenge is to move away from low-cost production to higher added value through non-price factors such as design and quality. For example, the UK women's footwear industry has suffered massive import penetration over the past decade, as shown in table 2.2.

Year	Percentage of imports
1984	60
1985	59
1986	58
1987	63
1988	65

Source: British Footwear Manufacturer's Federation

Table 2.2 UK import penetration in women's shoes

In order to combat this, recommendations from a recent consultant's report suggest the need to :

- develop niche products

- improve design and quality

- acquire a quick response capability, especially in order to deal with repeat orders (there is a potential competitive advantge here since orders from Latin America can take four months and from Taiwan up to six months)[24]

Similarly, in the furniture industry, where the UK market is worth nearly £6 billion, massive import penetration has already taken place, with 1989 imports running at three times exports in terms of value. Yet there are opportunities, as another consultant's report suggests:

'there is massive scope for the more aggressive retailers to create a competitive edge through distribution, individuality and product style.'[25]

In the field of textile production, Japanese industry suffered serious setbacks due to the high value of the yen, especially in relation to Taiwan and South Korea as lower-cost producers. The resulting strategy was to move away from price competition and into higher added value and product development, stressing better performance: there has been a decline in volume, from around 26 million square metres in 1986 to 20 million in 1988, but their value has held steady at some ¥5 billion per year.[26] This is becoming a worldwide trend; as Rawsthorn comments:

> The structural changes in textile production have been accompanied by changes in the marketplace. The speed at which fashions change and the need to keep stock to a minimum mean that the most successful textile companies are those that offer the fastest and most flexible service to their customers.[27]

One example of this can be found in the hosiery business. The marketing of stockings, socks and tights has traditionally been on the basis of price. Indeed, during the 1970s such items became a supermarket product, sold on the basis of 'pile 'em high and sell 'em cheap' – often with the products offered in only one size and one colour. However, the industry has now been transformed into a successful fashion industry. In the UK it is worth some £400 million per year, and is characterized by the emergence of specialist retailers such as Sock Shop. Import penetration is very low compared to other segments of the clothing industry, at around 20 per cent, and the industry's recent growth has been in value rather than volume (five times that of volume), reflecting the changed emphasis on quality, design and product innovation.[28]

Finally, in the clothing, footwear and leisure goods market the move has been towards 'designer-ism' in recent years, combining the frequent product innovation of such fashion markets with increasing customization and personal styling. The implications for manufacturing have been to emphasize the need for accurate information about demand and fashion trends and the essential feature of responsive and agile production systems.[29] For example, in a recent survey of the UK clothing industry, conducted by the Centre for Business Research,[30] respondents were asked to locate the position of their firms on a matrix of volume and variety of production. Even allowing for different interpretations, it was clear that the industry is moving towards placing much more emphasis on flexibility.

Small-batch producers

If the above changes suggest major problems for firms which have traditionally made high volumes in low variety, they have a familiar ring for those firms which traditionally produce in small batches. This is a large sector, including most capital equipment manufacturing (such as machine building of various types) and the engineering industry (for example, the majority of products made in the US engineering industry are in batches of less than 50). In such firms the traditional pattern has been one which involves offering a relatively high degree of variety to customers, and a flexible service in terms of meeting particular needs on product specification, delivery dates, production volumes, and so on.

The difficulty here is that manufacturing efficiently becomes something of a balancing act. The preferred picture from the customer side would involve more variety in the products offered, more customer-specific features, more flexibility in service and more frequent introduction of new products. But from the manufacturer's point of view, every increase in variety poses further problems and introduces more uncertainty into the manufacturing process.

Typically, a small-batch manufacturing operation in the engineering sector will involve a variety of machine tools. A product will visit a number of these machines in a sequence and, since they will need to be set up for that particular job, the product will be made in as large a batch as possible – the economic batch quantity. Thus, for much of the time, products in a batch are queuing with the rest of their batch for machine space, or waiting for the batch to be finished before moving to the next operation. The calculation of the economic batch quantity is designed to minimize the cost penalty of having stock tied up in these batches while keeping machines utilized and the customer happy (by making enough to satisfy his order).

As long as this system only involves a small variety of products, it works well. But as variety increases and machines have to be stopped and reset for different products, so the problems multiply and the possibility of accurate control over production as a whole diminishes.

Under these conditions even a simple problem such as a machine breakdown becomes a source of major difficulty since it adds to the queuing, (or it may happen when a batch is half finished, for example). It becomes very diffiicult to provide customer service under these unpredictable conditions – and the effect is to encourage customers to push manufacturers hard. Production thus moves from being a planned and manageable operation to one in which the loudest customer voice at any one time gets the attention; but of course, rescheduling a particular batch to pacify one customer further disrupts the rest of production. Batches jump queues and the disruption gathers momentum. With so much half-finished stock around it becomes possible that some batches will be inadvertently overlooked – until customer pressure becomes unbearable – and the cycle continues, moving progressively further from planned control flow towards chaos and crisis.

This may sound melodramatic, but it is a pretty accurate reflection of the situation in many small batch production facilities. Ideally, managers would like to reduce variety and increase batch size, since this would help them to reduce the complexity to manageable levels. However, as we have already seen, the market wants just the opposite – more variety, more customer service, more responsive delivery and so on. As an example, the shortening product life cycles (measured in years) in various branches of Japanese engineering are shown in table 2.3.

Sector	1981	1984
General machinery	4.979	4.03
Electric machinery	5.443	4.451
Transport machinery	4.72	4.206
Precision machinery	4.981	3.865
Source: JSPMI/ERI, Mori, 1989		

Table 2.3   Product life cycles in Japanese engineering industry

Jobbing

The final category of manufacturer which we should look at is the small firm specializing in making to meet a particular customer order. Here the emphasis has always been on providing a high level of customer service and support; on being able to make to highly individual specifications, meet particular quality and delivery requirements, and so on. However, it is becoming increasingly difficult to pass on the high costs of doing this to the customer in the form of higher prices. One option is to move the firm further up the service chain, developing the capability to offer more than simply a manufacturing capability by moving into research and development, design and other areas. A second – and much more limited – approach is to look for ways of achieving higher levels of productivity within what still needs to be a highly flexible and responsive operation.

2.6   PROBLEMS ACROSS THE BOARD

The pattern described above suggests that – right across the manufacturing spectrum – mass markets are being replaced or at least modified by more highly segmented ones as we move into the 1990s. The overall volumes may be the same but are now made up of many different 'packets' of much more specific demand. The emphasis begins to shift towards what have been called 'economies of scope' rather than just of scale; that is, how to make things in smaller amounts and with more frequent changeover, without incurring cost penalties in their manufacture.

Such growing fragmentation and segmentation of the market need not be seen simply as threats to established manufacturers: these trends also open up opportunities, particularly for smaller firms and nations, to occupy market niches. Some commentators argue strongly for new options in economic growth that are made possible by pursuing a policy of 'flexible specialization'; that is, becoming good at responding to the lower-volume, high-variety kind of market environment.[31]

2.7 THE STRATEGIC RESPONSE

Manufacturing strategy is shaped by both the internal pressures and the market environment in which firms operate. The key competitive priorities amongst manufacturers in Europe, Japan and the US (based on a regular survey of senior manufacturing executives in those countries) are listed in table 2.4, highlighting clearly the growing importance of non-price factors. This information refers to the major advanced nations, but the picture in the developing world, or in the major centrally planned economies is essentially similar. If anything, the concerns are even more urgent: not only is there a need to make more efficient use of often scarce resources of capital, skilled labour and so on, but there is also considerable pressure to achieve a level of international competitiveness. These concerns become of critical importance as such countries move

Europe/US	Japan
Consistent quality	Low prices
High-performance products	Rapid design changes
Dependable deliveries	Consistent quality
Fast deliveries	Dependable deliveries
Low prices	Rapid volume changes
Rapid design changes	High-performance products
After-sales service	Fast delivery
Rapid volume changes	After-sales service
Source: INSEAD, 1986	

Table 2.4   Competitive priorities into the 1990s

towards open market-based economies. As we have already seen, this will increasingly depend on strategies which exploit non-price factors: although a handful of developing countries may be able to exploit local cost advantages to become low-cost mass producers, this option is limited. Since producers in industrialized countries are continually trying to improve their operations through the use of technology, the price advantages to countries with low materials, energy or labour costs are likely to be whittled away. At the same time the number of countries trying to compete on this basis makes it a precarious strategy on which to base long term development.

Some countries have tried alternative approaches, notably on the Pacific Rim where nations such as the Republic of Korea have adopted an export-led approach, first exploiting their local cost advantages but increasingly moving to compete on design, quality and delivery. Others, such as many Latin American nations, stand at something of a crossroads, needing to shift their emphasis on manufacturing strategy to reflect these new realities. Their need is for more indigenous design, for better quality, for more reliable delivery, for service – and for flexibility to help them to make better use of often under-utilized industrial capacity. And all of this has often to be achieved with little new capital investment, because of shortages of hard currency and the crippling pressure of debt repayment.

In those countries that have been characterized up to recent times by centrally planned economic policy the pattern in manufacturing has been a systematic development of the mass production concepts which we saw earlier. It is here that giant, highly integrated plants, geared to the production of a narrow range of products, were to be found. Since labour-saving is not a dominant innovation motive, there has been a tendency to develop labour-intensive production, carrying the principles of division of labour and specialization to their limits. Such enterprises have also lacked the stimulus of competition in the past, and this has diffused throughout the organization, reducing the motivation to improve or to innovate. But now, in the wake of 'perestroika', a growing awareness is developing of the urgent need to improve manufacturing along several of our key non-price dimensions. Quality, variety, flexibility, design and delivery are all becoming as important as output, particularly in those industries which are attempting to compete in international rather than closed domestic markets.

2.8 BEYOND FORDISM

As we saw, the powerful influence of Henry Ford in establishing a blueprint for subsequent manufacturing organizations is reflected in the label 'Fordism', given by some economic historians to the approach. It is recognized that Fordism was about much more than just building cars: it represented a coming together of several ideas dating back over the previous two centuries, regarding not just the organization and management of production but also the basis of political economy.[32]

As a consequence of its dramatic impact on productivity in an era of demand for high volumes of new products, Fordism became the dominant institutional form with which the early twentieth century system operated. However, by the mid-1970s what many commentators have called a 'crisis of Fordism' developed, in which the model seemed increasingly inappropriate. Amongst features of the environment which had changed were the saturation of mass markets, the emergence of rigidities in the labour process (especially a wage explosion resulting from strong unions and non-compliant labour) and the rise of the NICs, bringing more competitive actors into the picture.

In the 1980s we have seen the emergence of what are sometimes called neo- or post-Fordist models, suggesting that the mismatch between system and environment has reached the point where a new model is needed. Such models for best practice are by no means clearly defined, but they are likely to be characterized by common elements such as market fragmentation, flexible specialization and an emphasis on flexibility – in production, in labour, in inter-institutional relationships and beyond.[33]

2.9   RESPONDING TO THE CHALLENGE

There is growing recognition of the need for alternative approaches in response to the manufacturing environment of the 1990s. Such responses need to offer a resolution of the apparent trade-off between productivity and flexibility, quality and responsiveness. A radical change in the technologies and organization of enterprises is likely to be needed to keep up with a rapidly changing and highly competitive market environment, while also maintaining productivity growth within the various manufacturing operations. This is not an uncommon experience: if we look back over the history of manufacturing we can see that the name of the game has changed at regular intervals, as discussed earlier.

As the manufacturing environment becomes increasingly turbulent and complex, the manufacturers who survive these changes will do so by making similarly dramatic changes in their own operations. They are looking for a *strategic* response to these challenges, not just a tinkering around the edges of the problem. One response is, of course, to do nothing, in the mistaken belief that the world has not fundamentally changed: there is a wealth of evidence to suggest that this is a recipe for disaster. Indeed, manufacturing history is littered with cases of firms or even whole industries which failed to recognize challenges and to adapt. Examples might include the failure of many Western shipbuilders to adopt the new design and assembly techniques which Japan and other Far Eastern nations implemented in the 1950s, the major threats to the motor vehicle and consumer electronics industries which Japanese production methods highlighted in the 1970s, and the near extinction of the Swiss watch industry in the face of microelectronics.

Another view is that manufacturing no longer matters and that future economic growth will be derived from the development of service industries. This view is flawed for at least two reasons. First, there will always be a need for wealth creation, not least to fund those public services such as education, welfare and health care, which may not necessarily be profitable but are nonetheless socially desirable. It is unlikely that even advanced service-based economies could create sufficient wealth to meet this need.

Second, such pessimistic views are often based on the assumption of a gradual shift towards a 'post-industrial' society. The implicit view is that manufacturing has a life cycle which cannot be reversed, that industries mature and then become uncompetitive in world terms and should be abandoned. But there is widespread evidence to contradict this view – manufacturing industry can and does show continuing renewal and renaissance. What has often been lacking is a sense of its strategic importance, and the managerial competence to make effective use of the available weaponry to implement such manufacturing strategy. Perhaps the most powerful of these weapons is technology, to which we now turn our attention.

Notes

1   For a discussion of these theories of innovation, see R. Coombs, V. Walsh and P. Saviotti, *Economics and Technological Change* (Macmillan, London, 1987), and R. Rothwell and W. Zegveld, *Reindustrialisation and Technology* (Longman, London, 1985).

2   For a detailed discussion of this, see J. Fleck, *The Development of Information Integration: Beyond CIM?*, PICT Working Paper 9, Edinburgh University, 1988.

3   For a detailed discussion of business strategy, and the techniques for analysis, see, for example, M. Porter, *Competitive Strategy* (Free Press, New York, 1980).

4   T. Hill, *Manufacturing Strategy* (Macmillan, London, 1985) .

5   F. W. Taylor, *The Principles of Scientific Management* (1911, reprinted Harper and Row, New York, 1949).

6   R. Jaikumar, *From Filing and Fitting to Flexible Manufacturing*, Working Paper, WP 88-045, Harvard Business School, 1988.

7   Jaikumar, *From Filing and Fitting*, p.35.

8   Taylor, *Principles*.

9   For a detailed description of the evolution of the Ford system, see H. Ford, *Today and Tomorrow* (Productivity Press, Cambridge, Mass., 1988 reprint) and W. Abernathy, *The Productivity Dilemma: Roadblock to Innovation in the Automobile Industry* (Johns Hopkins University Press, Baltimore, 1977).

10 J. Tidd, *Flexible Manufacturing Technology and International Competitiveness*, (Frances Pinter, London, 1991, forthcoming).

11 D. Altschuler, D. Roos, D. Jones and J. Womack, *The Future of the Automobile* (MIT Press, Boston, Mass., 1985).

12 *Automation* magazine, March 1989.

13 Abernathy, *The Productivity Dilemma*.

14   OECD, *Economic Survey of the UK* (Organisation for Economic Co-operation and Development, Paris, 1982).
15   N. Garnett, 'Forging a future from recovery', *Financial Times*, 12 September 1988.
16   *Financial Times*, 19 May 1989.
17   J. Ranta, *Issues and Problems of Flexible Paper-making*, Working paper, IIASA/Technical Research Centre, Helsinki, Finland, 1989 (mimeo).
18   P. Marsh, 'Big supplier starts to think small', *Financial Times*, 9 May 1989.
19   Yamauchi, quoted in J. Tidd, 'Next steps in assembly automation', *Proceedings of International Motor Vehicle Program. Policy Forum*, Acapulco, May 1989 (IVMP/MIT, Cambridge, Mass.).
20   'Japan's manufacturing industry retools at record rate', *Financial Times*, 22 November 1989.
21   'Japan's car industry changes to higher gear', *Financial Times*, 15 January 1990.
22   S. Milne, 'New forms of manufacturing and their spatial implications', paper presented to a conference on *Applications of New Technologies in Established Industries – Prospects for Regional Industrial Regeneration and Employment in Europe*, Centre for Urban and Regional Development Studies, University of Newcastle upon Tyne, 23–25 March 1988.
23   National Economic Development Office, *Advanced Manufacturing in Electronics*, NEDO Discussion paper, Electronics EDC, London, 1985.
24   TMS consultant's report, cited in 'Britain's slow shoe shuffle', *Financial Times*, 27 July 1989.
25   Verdict Consultants, reported in *Financial Times*, 29 August 1989.
26   *Financial Times*, 14 August 1989.
27   A. Rawsthorn, 'Woven to suit the changing market', *Financial Times*, 26 September 1989.
28   A. Rawsthorn, 'Price is not the sole criterion', *Financial Times*, 22 September 1988.
29   NEDO, *Dynamic Response*, (NEDO Books, London, 1987).
30   M. Whitaker, H. Rush and W. Haywood, *Technical Change in the British Clothing Industry* (Centre for Business Research, Brighton Polytechnic, 1988).
31   M. Piore and C. Sabel, *The Second Industrial Divide: Possibilities for Prosperity* (Basic Books, New York, 1984).
32   For a detailed discussion of this, see R. Kaplinsky, ' Restructuring the capitalist labour process: some lessons from the automobile industry', *Cambridge Journal of Economics*, (1989).
33   Piore and Sabel, *Second Industrial Divide*.

3 Technology: the New White Knight?

3.1 INTRODUCTION

As was suggested in the last chapter, technology is not a magic wand which can be waved over the factory to solve manufacturing problems. It is the product of interacting forces which shape and mould it to meet particular needs and open up new opportunities. It is also important to see it not as a single product such as a washing powder, but rather as a field of opportunity. And within this field there are all sorts of options to suit different users and to meet different needs. At the strategic level this is a critical point, since it implies that one key component of the effective management of technology will be the ability to make informed choices from within this field.

Before we consider some specific technologies, it will be useful to look briefly at some of the key trends which characterize the current generation of advanced manufacturing technology. In particular, we will focus on three: the role of information technology, the trend towards integration in manufacturing activities, and the massive growth in the supply side for advanced manufacturing technology.

3.2 THE ROLE OF INFORMATION TECHNOLOGY

These days the term 'advanced manufacturing technology' is often synonymous with information technology (IT). Yet even ten years ago the level of application of computers and related technology was relatively low. Although industrial computer systems were available in the 1960s, it was not until the mid-1970s, with the widespread diffusion of microprocessor control, that things really started to happen.

During the past ten years the transformation has been dramatic. Estimates suggests that around 70 per cent of all factories in industrialized countries are using IT in some application, and within larger firms or sectors, such as the electrical or chemical industries, the figure is closer to 100 per cent. And if we look at the types of application, the extent of this penetration becomes even more marked: almost all manufacturing activities, from initial design and pre-production planning work right through to final delivery, can employ some aspect of IT.

Many commentators see in IT the elements of a major technological revolution. Unlike many technologies which are specific to a particular process or area in manufacturing, IT is a pervasive technology, 'able to reach the parts other technologies cannot reach', what Freeman calls a 'heartland' technology. [1]

At one level this pervasiveness is easy to understand. IT is usually defined as the convergence of computers and communication technologies, but for our purposes it is more useful if we think of it not in terms of what it is, but of what it *does*. IT makes possible radical improvements in the way in which we store and retrieve information, the way we process it and the way in which we communicate it.

If we look at a typical factory and break down the activities going on within it, then we find that the majority are, at heart, information activities. All the production planning and control processes, from sales order processing, through purchasing, capacity planning, production scheduling, transport, quality, right through to final despatch, are essentially information-based. Monitoring and control of the various process variables involves combinations of information recording and processing (figure 3.1). Design is increasingly an information-based activity. Financial controls of various kinds are critically dependent on information flows and stores.

Without such information manufacturing deteriorates into a random and anarchic process which cannot be planned or controlled, and which produces wide variations in quality. The need for a technology to assist with this activity has long been present and has pulled through some useful and ingenious solutions. But the emergence of a technology the primary focus of which is the improvement of information activities across a very wide range of applications has given an enormous boost in the area of monitoring.

For an IT-based system, such monitoring activities are simplicity itself. All that is required is a sensor of some kind, to detect in physical terms what is being measured and to translate that into electronic information, and a computer to handle the incoming information and to display and store it.

In practice, the basic process is so simple that a number of sophisticated features can be included to improve the overall management and presentation of the information collected. These options include printing, analysis of trends over time, different types of display (visual, aural, and so on), but in essence it is still the same basic and very simple activity. Such monitoring is at the heart of a wide range of industrial processes, applied wherever something has to be measured or checked.

However, although there is now an extensive range of applications based on monitoring and measuring accurately, rapidly and with sophisticated tools for analysis and display, this only scratches the surface of IT's potential impact on manufacturing. The next step up the ladder is to 'close the loop' and use the information collected to help control the process itself. This involves comparing the currently measured state with information about what it ought to be and – as a result of the comparison – to calculate whether control action needs to be taken, and if so what (figure 3.2).

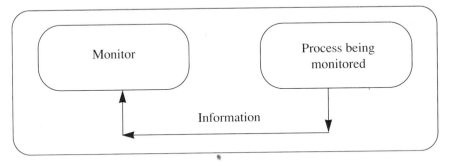

Figure 3.1   Simple monitoring loop

In a simple case, if we want to control a pan of water boiling on a stove, then we need some way of measuring the temperature (monitoring) and some idea of how hot we want the water to be. If the temperature of the

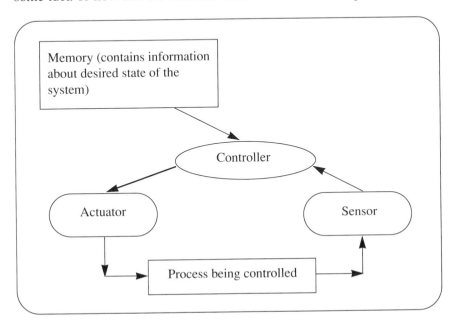

Figure 3.2   Simple control loop

pan is too hot, we must take control action to bring it into line with what it should be – by turning down the stove. If it is too cold, we turn up the stove; and if it is correct, we do nothing except keep on checking at regular intervals.

This is the basis of industrial process control. Whether we are talking about the speed of a single drill in a carpenter's shop or controlling the activities of people, materials, energy and so on flowing in and out of a whole factory, the same basic information principles apply. And, once again, this is an activity for which IT is well suited. Over the past 15 years microprocessor control loops have found their way into an enormous range of process control applications within all sectors of manufacturing. The list below, of control projects which have been carried out in the food industry, by one industrial automation firm using a standard microprocessor system gives some indication of the pervasiveness of this control technology. The applications are all based on the same basic family of microprocessors and demonstrate the high degree of flexibility of the technology in being tailored to suit particular requirements. All of these applications were introduced within a four-year period. Numbers in brackets indicate more than one project involving the same type of application:

- chocolate moulding control (5)
- soft drinks blending
- flour mill control (5)
- chocolate products manufacture (2)
- weighing and packaging control (3)
- keg filling (2)
- malting house control (2)
- batch mixing and continuous cooking
- refrigeration plant
- grain handling
- canning line
- tea blending
- boiler control
- bread mixing machine control (2)
- chocolate wrapping machine control
- monitoring and control (4)

- coffee process plant (2)

- tobacco handling control (2)

But even extensive use of IT in control loops does not represent the full capability of the technology. It can also be used to control several loops simultaneously and to handle those which are interdependent. And it can begin to combine information from different sources into an *integrated* control system; moving from providing benefits of improved efficiency – doing better that which was always done – to doing things in totally new ways that contribute to overall effectiveness. Some of this range of applications is set out in table 3.1.

Discrete application ◄——————————————► Integrated system		
Basic monitoring and measuring activities	Monitoring and control activities	Integrated monitoring and control activities
Production monitoring	Production control	Integrated production management systems such as MRP2
Process variable monitoring – speed, temperature, pressure, and so on	Process control of individual variables	Integrated process control for several variables simultaneously
Stock level monitoring	Stock control	Integrated inventory management
Energy usage monitoring	Energy-saving controllers	Integrated energy management systems
Stand-alone equipment – robots, CNC machine tools, and so on	Integrated production cells, for example linking robots, tools and handling	Flexible manufacturing systems and computer-integrated manufacturing
Computer-aided draughting	Computer-aided design	CAD/CAM, computer-aided design and manufacturing

Table 3.1 Choice in level of application

3.3 CHOICES IN THE APPLICATION OF IT

Another feature of IT which encourages its diffusion into so many nooks and crannies of manufacturing industry is the range of choices which it offers in the way in which it can be applied. This runs from simple incremental improvements, which substitute for that which has always been done, right through to radical new processes. But it also allows for wide differences in the extent to which the user becomes involved in the design and programming of the system.

For many users with relatively standard applications it will usually be easiest to buy something ready-made 'off-the-shelf': indeed, most modern equipment now comes with microprocessor-based control systems. There may be a need to tailor the technology to particular requirements by programming, but this is largely a matter of adapting a standard solution for a common application. In some cases – for example, in the process industries – standard control loop controllers can be purchased off the shelf, and the user has only to specify the relevant parameters to be controlled and program the system.

But as we move towards more specific and specialized applications, so the need emerges for more active involvement by users. In many cases the lack of a large enough market for a standard product will mean that there is little or no choice of equipment 'off-the-shelf' and so users have to become involved in developing solutions for themselves. Here the power of IT as a general-purpose control technology emerges. As long as there are suitable sensors and actuators and the process is understood (in terms of the key control parameters), it is possible to develop a controller for any application.

Nor does the application always have to involve a new piece of hardware. Many successful applications of IT have involved a process known as 'retrofitting'; that is, fitting a modern control system to an existing piece of equipment. An example might be the refurbishment of machine tools, replacing the old control systems with some form of computer-numerical control. This option is of particular value to organizations in which there are constraints on the purchase of new equipment, or where there is no suitable new technology available.

For example, the UK firm of Sturmey Archer used a variety of old equipment to produce various bicycle parts. Its plating plant was a constant source of problems due to the unreliability of its relay control system – on one occasion it took a fitter two weeks to replace burnt-out cable and faulty switches. The system was replaced by a simple microprocessor control unit costing less than £500, which gave much better reliability and also rapid fault-finding when problems did occur. The firm subsequently fitted their remaining two plating lines with similar controllers and then began looking for other applications. They drew up a 20-machine list, based on an analysis of the frequency and cost of breakdowns and the importance to production of that piece of equipment. Microelectronics was able to make a major contribution here. For example, a swarf dock (a controlled lift for transporting swarf and waste) had major reliability problems due to clogging with swarf.

Repairs took place on a daily basis, but when the controls were replaced with a microelectronics unit (costing £480) the lift was able to work for weeks without attention. Another application was a six-station drill, which used limit switches to position a turntable, and was subject to regular breakdown due to faults in the gearing and switches; again, the use of microprocessor control improved the accuracy and reliability considerably. Finally, a 40-year-old broach machine was retrofitted with a programmable controller which, once again, reduced downtime and improved the quality and flexibility of control.

Significantly, each of these applications was installed without external expertise, further underlining the relative ease with which micro-electroncis technology can be retrofitted to existing plant.[2]

3.4 DIFFUSION OF IT: A SOLUTION LOOKING FOR A PROBLEM?

IT has sometimes been called 'a solution looking for a problem' – and for management the question is no longer one of whether or not to adopt the technology but of what choice of system to make. What do we need? Can we obtain a standard off-the-shelf system or must we develop it ourselves?

Such a widespread choice means that there is almost no application which cannot benefit from IT – and the overall diffusion patterns support this view. Indications are that considerable investment is now going into IT in manufacturing: for example, revenues for computers in manufacturing in the US in 1987 were over $2.5 billion and are predicted to rise to over $4 billion by 1994.[3] UK data for the engineering industry in 1988 suggests that the overall value of investment in some 52 500 systems was around £2 billion.[4] Comparable data for the process industries indicates a further £700 million invested in around 28 000 computer systems.[5]

The diffusion of the technology in manufacturing processes in the UK is indicated in figure 3.3.[6] This suggests an overall level of use of around 70 per cent, which is similar to findings from the US for 1988 where (in a sample of 10 500 firms) 68.4 per cent of firms were using the technology[7] and for Canada in 1987 where 50 per cent of all firms used the technology in some form.[8]

In figure 3.4, drawn from several studies, it is indicated that this pattern was broadly similar in other industrialized nations in 1983, and there is no reason to expect any divergence from this trend in more recent years.[9]

Diffusion by sector

Applications of microelectronics exist in all industrial and service sector branches, but the emphasis varies considerably. In general, the electronics and electrical industries are the major users, followed by the

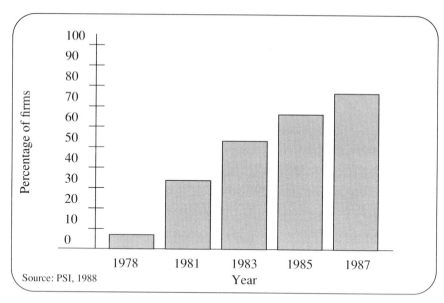

Figure 3.3   Diffusion of IT in UK manufacturing

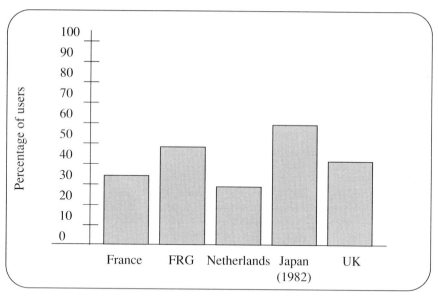

Figure 3.4   International diffusion of IT, 1983

capital-intensive process industries and those sectors (such as food and drink and some branches of mechanical engineering) in which discrete applications (such as weighing or measuring) are important. Sectors which have traditionally been labour-intensive or which are concerned with raw materials that are difficult to manipulate in a predictable fashion

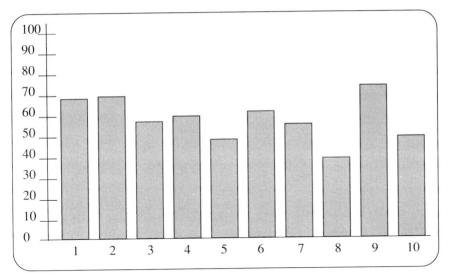

Figure 3.5   Diffusion of IT by sector in the UK, 1987. 1, Food and drink; 2, chemicals, metals; 3, mechanical engineering; 4, electrical engineering; 5, vehicles; 6, metal goods; 7, textiles; 8, clothing; 9, paper and print; 10, other. *Source:* **PSI, 1988**

– such as clothing or footwear – have been much slower in their take-up of microelectronics. The sectoral breakdown is indicated in figure 3.5.

Data from the UK *Engineering Computers* magazine survey, covering the engineering sector only, suggests an overall level of usage for 1988 of 88 per cent, with a prediction that this will rise to 95 per cent of all engineering firms.[10] US figures from the Bureau of the Census also confirm this level of use in engineering and related industries.[11] In UK data on the process industries the average varies from 67 per cent for smaller firms (less than 200 employees) up to 77 per cent (less than 500) and 91 per cent for plants employing over 500.[12]

Diffusion by size of firm

It might be assumed that technical complexity would make micro-electronics the province of larger firms but, while it is true that larger firms do have more resources to devote to such innovation, small firms have succeeded in applying microelectronics very successfully across a wide range of sectors and in both products and processes. Again, that is high-lighted in figures 3.6 and 3.7, which show data for the UK and US (1987/8) and comparative data for several industrial nations (1983).[13] Canadian data for 1987 also suggests that 62 per cent of large firms and 36 per cent of medium-sized firms were using the technology.[14] Data for the engineering and process industries in the UK (1988) confirms this distribution.[15]

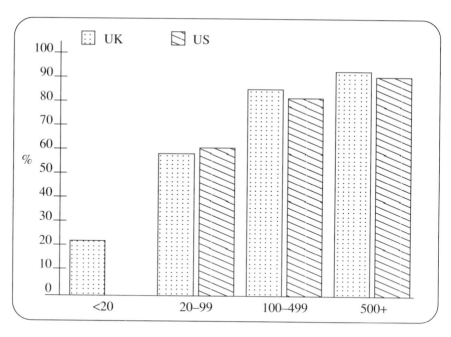

Figure 3.6   Diffusion of IT by firm size in the UK and US, 1987

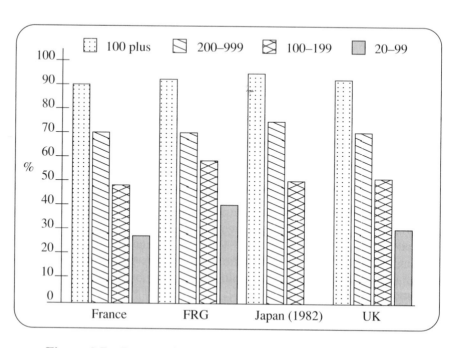

Figure 3.7   International diffusion of IT by firm size, 1983

3.5 EXPERIENCE WITH IT: THE PROMISE

Apart from its near-universal applicability, one reason why the technology has diffused so widely is that it can be used for so many different reasons, depending on circumstances. Whereas most technologies offer benefits within a narrow range – improved speed, savings in key inputs, improved quality, and so on – IT can offer benefits across a very wide range. Data for the UK, for example, confirms this in the list of benefits which firms report for the technology, listed in figure 3.8:

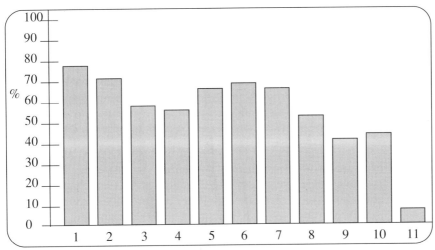

Figure 3.8   Benefits from the use of IT in the UK, 1987. 1, More consistent, better quality product; 2, better control of production processes; 3, greater speed of output; 4, more flexibility in production; 5, lower production costs; 6, more efficient use of labour; 7, more efficient use of equipment; 8, more efficient use of materials; 9, more efficient use of energy; 10, improved working conditions; 11, equipment more reliable, easier to maintain.
Source: **PSI, 1988**

This means that, even for firms in the same sector, use of IT may be for different motives and in different applications, supporting the view held by many that it is much more than a single technology: like steam power or electricity before it, it is a *generic* technology. And it is this property – the potential to offer a wide range of competitiveness-enhancing benefits right across the industrial spectrum – which really explains the claims for its revolutionary status.

3.6 EXPERIENCE WITH IT: THE REALITY

In practice, the more extreme claims about the speed and impact of IT have not, so far, been borne out. The 'revolution' has taken a lot longer to work through than was originally thought. Although diffusion is now at the 70 per cent level in most factories, this is almost 20 years after the invention of the microprocessor and 30 years since computers began to be used to control manufacturing operations and processes. And although the threat to non-users of IT continues to loom on the horizon, there is little indication yet that the widespread collapse of these industries has taken place.

What has happened is a gradual *evolution*, a learning process in which a growing number of firms have begun to make use of the technology and have begun to reap some of the very significant benefits. Examples abound, and the following list is merely to convey some of the flavour:

1   In the engineering sector a UK firm, Johnston Engineering, spent £1.5 million on CNC turning centres and related technology to facilitate their production of motor components. This investment has helped them to reduce lead times from 5.3 days to 1 day and to cut the work-in-progress inventory. Another components supplier, Retrac Products, were able to cut their lead time on car bumper production from 10–12 weeks from receipt of order to 12 days.[16]

2   In the clothing industry IT has been extensively used at all stages, from pre-production through assembly to finishing. In the pre-assembly stage use of computer-aided design (see chapter 7 for a fuller description of this technology) has produced savings ranging from 5 per cent to 12 per cent on materials and an increase in flexibility as a result of reduced times for grading and marking (cut by a factor of between two and six). The use of automated cutting equipment has also been reported to reduce direct labour by as much as 50–60 per cent, although improved material utilization is the principal reason usually cited for adoption, with savings of around 5 per cent.[17] In assembly technology, computerized sewing machines of various types have had a major impact on the industry. Pre-programmed dedicated units, designed for high-volume, low-variety work, have been responsible for increases in labour productivity between 38 per cent and 75 per cent depending on the particular activity and machine. Similar figures can be achieved from pre-programmed convertible units which are designed for some reprogrammability to enhance their flexibility, while fully operator-programmable units offer high flexibility and increases in labour productivity of between 18 per cent and 46 per cent for individual activities. [18] In the handling and finishing area a wide range of activities have been affected by IT, including pressing and finishing, handling and overall production management and control.[19]

3   In the food industry a wide range of applications can be identified. For example, a UK dairy (Cliffords Dairies) spent £2.75 million on a new bottling plant to handle 290 000 litres of milk daily. The whole system (four milk reception pumps, two chillers, seven

storage silos, two pasteurizers and four finished milk tanks) is under microprocessor control. The system monitors 350 separate variables and directly controls 260 of them. Benefits include labour saving, space saving, materials saving and increased flexibility.[20]

4   In the brewing industry microelectronics is utilized in the control of brewing (from mashing to conditioning), in addition to applications in packaging and stock management.[21]

5   In another food industry example, a factory for the production of frozen pizzas and other comminuted meat products, IT is used to support every stage of the manufacturing process. In the meat area, frozen blocks are delivered, weighed using microprocessor-controlled scales, checked for metal using an electronic detector and stored in a cold store. A visual image analyser digitizes photographs of the meat and a microprocessor calculates the fat : lean ratio. This information is then used to calculate the formulations to be used in later processing. Prior to processing the meat is allowed to thaw over a controlled 20-hour temperature cycle: this stage is managed by a PLC system. The route from here on is designed as an automatic flowline, but through the use of computer controls and layout planning it is possible to retain a high flexibility in product range. In essence, frozen meat goes in at one end and a high variety of meat products emerge at the other. Hamburger patties, for example, emerge in this way and are then weighed and quality checked (including further metal detection and optical scanning) under computer control. They are then fast-frozen for 20 minutes in another continuous flowline which leads to the packaging area. Once again, computer controls on advanced machinery mean that any one of five packaging lines can handle any product, giving further flexibility to the plant. In the bakery side of the operation a similar pattern is found, with computer-controlled dough mixing and continuous flowline shaping and cooking, with advanced energy-saving features. Over the entire process, information for production management is obtained from each manufacturing station and passed to a production management computer system.[22]

6   In a survey carried out in 1980 into the use by small-scale firms of CNC technology, the main motives were found to be overcoming shortages of skilled labour, saving on labour costs, coping with increased demand for precision, and increasing flexibility to handle a greater number of smaller batches. Significantly, the entrepreneurs interviewed argued that they would have been unable to obtain attractive work without CNC technology.[23]

7   In Hong Kong, textiles and garments are amongst the oldest industries. Many of the enterprises are small and face problems of

low productivity, but microelectronics can increase this in a number of ways. A low-cost CAD/CAM system has been developed by the Hong Kong Productivity Association, allowing users to grade and plot patterns quickly and cheaply. Benefits from this and other applications of microelectronics include the reduction or elimination of production bottlenecks, reductions in material costs through timely feedback on usage levels, and a general increase in production and financial efficiency.[24]

8   In a survey of 19 small and medium-sized enterprises in Brazil on the use of CNC, 17 of the firms had plans for further adoption of the technology on the basis of their successful experience so far. The main motives for adoption were to improve quality and flexibility, to produce to higher tolerances with consistency, and to raise productivity.[25]

3.7   THE TREND TOWARDS INTEGRATION

Although IT has undoubtedly been a powerful force for change, it would be wrong to see this as the only factor which has influenced the emergence of advanced manufacturing technology in the form available to us today. Perhaps the most significant trend – which IT has accelerated, but which it predates by many years – is that towards *integration*.

Innovation in manufacturing has been taking place ever since the Stone Age, as new and better ways of doing things have been discovered and developed. But there is a qualitative difference between innovations which are primarily associated with 'doing what we've always done, but a little better' and those which fundamentally change the nature of the process to which they are applied. In the former case we are really talking about a *substitution* process: a typical example here would be the replacement of a machine with one designed to work faster or more accurately. In the latter case the machine could be radically extended in its capability; for example, by making it able to perform several functions instead of one, or by adding an 'intelligent' controller.

The distinction is one of *integration*. Through a synthesis of different elements the whole becomes greater than the sum of its parts. As we move from substitution towards more integrated forms, so we bring together more of the previously separate functions in the manufacturing process. At the same time, the benefits which the technology offers increase with higher levels of integration. We move from 'more of the same but a little better' – faster, more accurate, and so on – to radically new opportunities which offer significant improvements across a broad front in quality, flexibility, productivity and so on. In terms of systems theory, more integrated systems have 'emergent properties', appearing only at the higher levels of integration of subsystems.

We can see this in a number of examples. In the field of manufacturing, for example, the earliest machines built upon the integration of craftsman's tools with new sources of motive power. Subsequent development enabled a single machine to perform multiple functions, so that although its cost and complexity rose, it replaced several older-generation machines. Nor was this process confined to the physical technology alone. As we have seen, the process of organizing production and the use of labour within a pattern of work organization was also gradually integrated in such a fashion that, by the 1920s, the Ford lines represented the triumph of integration, linking men and machines into a complex but – even by today's standards – an extremely efficient integrated system. This also involved the integration of supply chains and distribution.

Other developments in machinery in more recent years have included the integration of 'intelligence', gradually substituting the judgement of individual craftsmen by the incorporation of some form of machine controller: originally NC, then CNC/DNC, then FMS, and now artifical intelligence. Over a period of 150 years or so we have moved via this process of integration from a single machine tool to a complete manufacturing system. On the way, the range of benefits emerging from the innovation have moved from better and faster machining to much more powerful and strategic advantages such as lead time reduction, quality improvement, flexibility improvement and inventory saving.

The same trend can be observed in other areas of manufacturing. For example, in the field of design the various tasks associated with this complex process have gradually become integrated into computer-aided design systems. In this process the traditional drawing office, with its rows of draughtsmen crouched over drawing boards, has been replaced in many cases by a computer-aided design facility in which the designers work with computer terminals. This changes the nature of their work a great deal, even though they are superficially still carrying out the same task, except with an electronic pencil.

The first change is that the activities of design and draughting have converged. Whereas before much of the design department's work was laborious drawing out of ideas, and subsequent redrawing to accommodate changes and improvements, this can all now be done using the computer system. More importantly, the drawing on which a team of designers are working at any moment is automatically updated with the results of changes made by any other designer. This means that, for the first time, all designers are working on the same project and, in the design of complex assemblies such as motor cars, such a feature can mean significant savings.

However, even allowing for the very powerful contribution of computer-aided design to the draughting and design process, its real

significance emerges as IT facilitates its integration with the manufacturing process itself. Since CAD systems make use of information
coded in electronic form, it follows that other systems – such as those for
computer-aided manufacturing, which also use such information – can be
linked in via some form of network. This is the basis of CAD/CAM
(computer-aided design and manufacture) in which not only can the
product be designed on a computer screen but, when the design is finally
refined, the necessary instructions can be generated and sent to the
machine tools and other devices which will actually manufacture it.

The advantages of this kind of integration are enormous, and extend
beyond the generation of designs and the relevant information necessary
for controlling the manufacturing process to those for other activities; for
example, managing the various coordination and control activities such
as the planning of material requirements and capacity planning, and
quality control. Benefits arising from this include significantly reduced
lead times, improved quality, better machine utilization and much
improved customer service.

One other feature worthy of comment here is that whereas Ford was
able to achieve very high levels of productivity and efficiency in his plant
through the integration of different physical elements into the assembly
line, this led to rigidity and centralization, which ultimately acted against
the company's competitiveness. By contrast – as a consequence of
network technology – today's systems can offer the possibility of *flexible
integration*, in which systems can be simultaneously tightly coupled and
highly centralized and also highly decentralized and autonomous.

3.8   COMPUTER-INTEGRATED MANUFACTURING AND BEYOND

Up to now, such integration has taken place within particular functional
areas of manufacturing. Therefore, the major changes in design have
remained largely in the areas of design and drawing. Improvements in
machining systems have mostly been confined to the factory floor, and to
specific areas within that. However, recent developments have begun to
blur the lines between functions – as in CAD/CAM systems, for
example, which make use of the feature that IT-based systems use a
common electronic language and can thus be configured to communicate
with each other via some form of network. This is the basis of computer-
integrated manufacturing (CIM) which can be defined as: 'the integration
of computer based monitoring and control of all aspects of the
manufacturing process, drawing on a common database and communicating via some form of computer network'.

This development is shown in figure 3.9 – from integration within
functional areas towards inter-area integration and, finally, CIM.

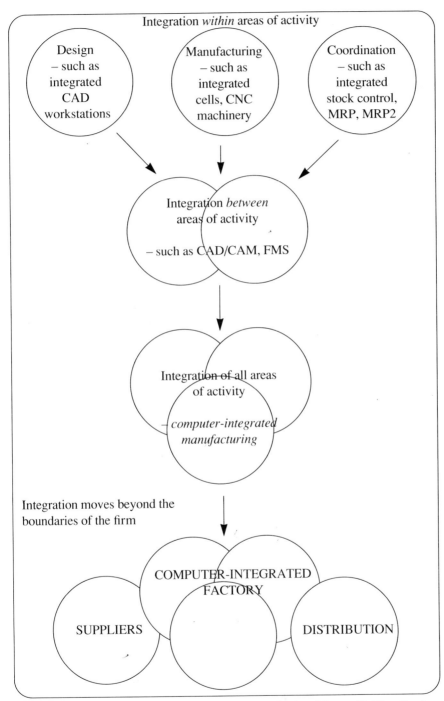

Figure 3.9 **Convergence to CIM.** *Source:* derived from R. Kaplinsky, *Automation: the Technology and Society* (Longman, London, 1984)

Of course, such integration does not stop at the boundaries of the firm. Integration via electronic means can also extend backwards along the supply chain (with, for example, shared design processes or electronic components ordering linked to inventory management computers), or forwards into the distribution chain, using what is termed 'electronic data interchange' (EDI) to speed the flow of products to outlets while also minimizing the inventory held within the chain.

Such developments towards CIM do not just offer considerable improvements in traditional ways of making things. They also open up completely new and often highly integrated options. And the contribution which such changes can make to dealing with the problems of the market environment, as we move into the 1990s and beyond, are equally significant. Pressures on firms to be more flexible, to offer high quality, better customer service, and improved delivery performance, and to emphasize design and other non-price factors all pose major challenges to manufacturers, adding to the 'traditional' burden of ensuring the effective use of inputs of energy, materials, labour and capital. In this context, CIM is seen as a major and valuable competitive weapon.

Against this backdrop of the trend towards integration in manufacturing we can begin to map applications of advanced manufacturing technology in terms of the extent to which they are located on the substitution /integration continuum. Some examples of this are indicated in table 3.2.

Substitution ◄————————► Integration	
Computer-aided draughting	Computer-aided design and manufacture (CAD/CAM)
Stand-alone computer-controlled machine tools (CNC)	Integrated flexible manufacturing systems
Stock control computer	Manufacturing resourses planning (MRP2)

Table 3.2   Substitution and integration in advanced manufacturing technology

3.9   SUPPLY-SIDE GROWTH

The third trend which is of importance in considering the development of advanced manufacturing technology concerns the supply side. As a consequence of the growing pressures on manufacturers to find solutions to the strategic problems of the 1980s and 1990s, and the emergence of powerful heartland technologies with widespread potential, we have seen an explosion of growth in the industry concerned with supplying manufacturing innovation. The IT industry as a whole is expected to be worth around $600 billion by the end of the 1990s, and to have overtaken oil as the world's largest industry. Although this figure includes a major contribution from telecommunications, it should be borne in mind that much from the future development in IT in manufacturing is likely to involve networked systems within and between firms. Estimates for 1985 suggest that global expenditure on automation was around $38 billion, rising to $50 billion for 1989 and likely to exceed $90 billion by 1992. In the UK a recent survey suggested that some £2 billion, equivalent to 20 per cent of all manufacturing investment, was being spent on CIM products.[26] US estimates for 1988 suggest annual expenditure on computer systems integration of $17 billion.[27]

Within the automation industry the sectors concerned with supplying advanced manufacturing technology (AMT) have already demonstrated very rapid growth: typical annual sales growth has been in excess of 20 per cent for many products or systems. This has both positive and negative consequences for the potential user. On the plus side it means that there is a proliferation of choice. For any item of AMT which a firm might be considering buying, there is a wide and growing range of options from which to choose. This choice spreads across several dimensions, to suit firm-specific elements, such as cost, firm size, industrial sector, level of skill and previous experience. The effect of massive supply side growth has been to create a buyer's market, in which customers can demand high levels of service and support, and customization to meet their particular needs.

However, on the negative side, the expansion of the supply side into a large and highly competitive industry means that there is considerable pressure to sell. For inexperienced users there is the very real danger of being subjected to high-pressure sales techniques, which are not intended to solve a particular manufacturing problem but rather to shift a particular seller's set of boxes. Thus there is an important need for user firms to develop skills in the choice of technology to meet their particular needs.

The problem is exacerbated as we move along the continuum from substitution innovation to integration. As we begin to talk of major investments in manufacturing systems covering a number of functional areas rather than single items of equipment, so we begin to approach the limits of experience on both the supplier and the user side. For example, an engineering firm is likely to understand the problems involved in buying a new machine tool fairly well, and the supplier of those tools will attempt to address these concerns. But when the question is one of a flexible manufacturing system, which brings together machine tools, computers, robots, automatic transport systems and software, the problem becomes much greater than most firms have experienced before. The limits of their learning curve will have been reached, and their ability to manage the process of selection and implementation will be limited.

On the supply side the same problem is also emerging, as the move towards selling *systems* rather than individual items of equipment comes to the fore. The supplier of machine tools is suddenly challenged to become a much more broadly based supplier, requiring competence in computer hardware and software, robotics and automated transport technology, factory automation networks and so on. In the short term it is likely that his ability to provide a solution to the user firms' problems may also be limited by his own inexperience.

The effect of this is to introduce considerable uncertainty into the selection stage of the innovation process. A number of strategies have emerged to cope with this problem, ranging from delaying investment, until such time as the market matures, to the extensive use of external consultants and advisors. In this connection it is interesting to note the rapid expansion of the manufacturing technology and management consultancy industry and the particular interest in 'systems integration contracting'; that is, the provision of services analagous to those of a 'managing agent' in a major building project, taking on the responsibility for putting different contributions together and for bringing the project in on time and within budget.

On the supply side there is a recognition of the limits of the ability of any one firm – even amongst the largest vendors – to supply complete systems expertise and products and this has again led to several coping strategies. There have been a variety of mergers, acquisitions and 'strategic' alliances between firms, all designed to try to broaden the range of products, competence and experience which can be offered.

Perhaps the most interesting trend has been the emergence of early users of advanced manufacturing technology (especially in its more integrated forms) as suppliers of systems and expertise. Their argument here is that early users had to undergo considerable learning and development in order to solve their own problems – often including coping with an immature and disorganized supply side which was unable to meet their particular needs. Consequently, they have now accumulated a reservoir of knowledge and experience on the basis of 'learning by doing', which has considerable commercial value.

Overall, the picture of the supply side is one of enormous potential and choice, but also of continuing development and maturation. This makes the selection environment a highly complex and uncertain area – and emphasizes the importance of skills and competence in this area as a key feature in the successful implementation of AMT.

Notes

1 C. Freeman, J. Clark and L. Soete, *Unemployment and Technological Change*, (Frances Pinter, London, 1983).
2 'How to give new life to your old equipment', *Works Management*, December 1983.
3 Market Intelligence Research Corporation estimates, quoted in *Control Systems*, July/August 1988.
4 Benchmark Research, survey on behalf of *Engineering Computers* magazine, May 1989.
5 Benchmark Research, survey of the process industries, *Works Management* magazine, October 1988.
6 Compiled from J. Northcott, W. Knetsch and B. de Lestapis, *Microelectronics in Industry* (Policy Studies Institute, London, 1988), and G. Vickery and L. Blau, *Government Policies and the Diffusion of Microelectronics*, (OECD, Paris, 1989).
7 Data from *Manufacturing Technology 1988* (U.S. Department of Commerce, Bureau of the Census, Washington D.C. , May 1989).
8 *Survey of Manufacturing Technology*, Classification Systems Branch, Statistics Canada, report 88-03-15, 1988.
9 Northcott *Microelectronics*. Vickery and Blau, Government Policies, p.??
10 *Engineering Computers*/Benchmark Research – see note 4.
11 US Bureau of the Census, – see note 7.
12 *Works Management*/Benchmark Research, see note 5.
13 Compiled from Northcott et al., *Microelectronics*; and US Bureau of the Census 1989 data (see note 7).
14 Statistics Canada – see note 8.
15 *Engineering Computers* and *Works Management* surveys – see notes 4 and 5.
16 *Works Management*, December 1987.
17 K. Hoffman and H. Rush, *Microelectronics and the Clothing Industry* (Praeger, New York, 1988).
18 Hoffman and Rush, *Microelectronics*.
19 M. Whitaker, H. Rush and W. Haywood, *Technical Change and the Clothing Industry*, Occasional paper, Centre for Business Research, Brighton Business School, 1989.
20 *Food Processing*, November 1983
21 National Economic Development Office, *Microelectronics and the Brewing Industry*, (NEDO, London, 1980).
22 *Food Processing*, March 1984.
23 S. Watanabe, 'Numerically controlled machine tools in Japan', in *ATAS Bulletin*, Issue 2, United Nations Centre for Science and Technology for Development, New York, 1985.
24 A. Bhalla, D. James and Y. Stevens, *Blending of New and Traditional Technologies*, (Tycooly International, Dublin, 1984).

25   J. R. Tauile, 'Microelectronics and the internationalisation of the Brazilian automobile industry' in *Microelectronics, Automation and Employment in the Automobile Industry*, ed. S. Watanabe (John Wiley, Chichester, 1986).
26   A. T. Kearney Consultants, *Computer-integrated Manufacturing; Competitive Advantage or Technological Dead-end?* (London, 1989).
27   For example, it is estimated that the computer systems integration (CSI) business in the US is now worth $17 billion annually. See 'Andersen gears up for fight over CSI', *Wall Street Journal*, 4 January 1989.

4 A Marriage Made in Heaven?

4.1 A MARRIAGE IS ARRANGED . . .

We saw earlier the range of challenges facing manufacturers as they move into the 1990s, and particularly the stress placed on non-price factors. To what extent can advanced manufacturing technology (AMT) offer a means of dealing with these challenges? Some of the main problem areas, together with the sorts of benefits which are potentially offered by more integrated forms of AMT, are set out in table 4.1.

At first sight this appears to be a perfect marriage, one likely to lead to considerable improvements in overall manufacturing performance and competitiveness. Some of the evidence from early users seems to bear this view out, with regular reports of spectacular gains along many of the key dimensions of manufacturing competitiveness. But it is important to note that this 'marriage made in heaven' is not always successfully consummated – and in some cases it can lead to a long and rather messy divorce. Some examples follow:

1   In 1989 the UK was reported to be investing in CIM at a rate of nearly £2 billion/year, equivalent to 20 per cent of all capital expenditure in manufacturing. But up to a third of that money was potentially being wasted – integration has occurred only at a technical level and not at an overall business level. In particular:

> benefits on the whole have been disappointing with an achievement of 70 per cent of planned gains . . . CIM has not resolved the problems of quality and performance to schedules as anticipated . . . MRP has only managed to tidy up and enforce disciplines without achieving the 2 primary goals it claims to resolve i.e. inventory reduction and adherence to deadlines.[1]

2   In an earlier study of users of various types of advanced manufacturing technology, carried out for the British Institute of Management by Cranfield Institute of Technology, managers were asked to rate their investments in terms of their (subjective) views of the return to the firm. Their responses are given in table 4.2,

61

Main problem issues as seen by senior manufacturing executives in Europe	Potential contributions offered by CIM
Producing to high quality standards	Improvements in overall quality via automated inspection and testing, better production information and more accurate control of processes
High and rising overhead costs	Improvements in production information and shorter lead times, smoother flow and less need for supervision and progress chasing
High and rising material costs	Reduces inventories of raw materials, work-in-progress and finished goods
Introducing new products on schedule	CAD/CAM shortens design lead time – tighter control and flexible manufacturing smooths flow through plant and cuts door-to-door time
Poor sales forecasts	More responsive system can react quicker to information fluctuations – longer-term, integrated systems improve forecasting
Inability to deliver on time	Smoother and more predictable flow through design and manufacturing stages makes for more accurate delivery performance
Long production lead times	Flexible manufacturing techniques reduce set-up times and other interruptions so that products flow smoothly and faster through plant

Table 4.1   Computer-integrated manufacturing: a solution for the manufacturing problems of the 1990s?

from which it can be seen that many are dissatisfied or disappointed with such innovations.[2]

3   In a study of 33 firms using computer-aided production management (CAPM) systems, almost a third of those interviewed considered their systems to be failures or, at best, only partial

Technology	Zero to low pay-off (%)
CAD	46
CAM	46
FMS	67
Robots	76
MRP	19

Source: New and Myers, 1986

Table 4.2   Problems in AMT integration

successes. The report concluded that '. . . even advanced users . . . . . . are not getting the full benefits from their systems'.[3]

4   In studies of firms using CAD, Senker and Arnold found that it took two years on average for them to achieve the 'best practice' productivity gains potentially offered by the technology.[4]

5   In a research programme on users of advanced manufacturing technology, carried out on behalf of the UK National Economic Development Office, Burnes reports that:

> . . . of the 21 systems observed, 12 were operating satisfactorily but the other 9 were performing considerably below expectations. Indeed, the performance of one system was so bad that the company eventually scrapped it altogether . . . even in the 12 instances where satisfactory performance was achieved, in four cases there were major problems and long delays had to be experienced in bringing the systems up to expectations.[5]

6   Another UK study commented that:

> . . . only about 25% of CAD/CAM installations in the UK are considered a success. From the logistics perspective, the CIM track record is even more suspect; if we consider logistics is all about having the right resources at the right place and time then a substantial number of MRP2 installations can be considered failures.[6]

7   In recent evidence from the USSR concerning flexible automation, it is reported that:

> no less than a third of the 50,000 industrial robots produced between 1981 and 1985 had not performed even one hour's work. A sample inspection made by the People's Control

Committee of the USSR in 1985 showed that the annual return on introducing 600 robots, at a cost of more than 10 million roubles, was a mere 18,000 roubles.[7]

From the above it is clear that simply investing in AMT is no guarantee of its successful performance. In a few cases the problems have been so serious that the firm has been forced to abandon its investment altogether. A recent example of this was the UK bicycle manufacturer TI–Raleigh which '. . . installed an ambitious computerised manufacturing resources planning system to control inventory and speed throughput. The system was so complex that an untrained shop-floor took elaborate pains to circumvent it. Raleigh was almost bankrupted'.[8]

Similar experiences are beginning to emerge at several plants in the US belonging to General Motors, where GM have tried to use technology to regain their competitive edge. For example, their Hamtramck plant cost $500 million and involves the largest robot population in the US – 260 robots plus 50 AGVs – while Buick City in Flint Michigan, at a cost of $400 million, is designed to operate with only 30 per cent of the inventory of conventional plant, reflecting the high investment in flexible manufacturing technology. The $40 billion Saturn project, which aims to create a new marque rather than just a new car, is based at Spring Hill, Tennessee and was originally planned to use the ultimate in CIM technology. In order to facilitate these ventures GM invested in one of the largest and most successful software companies, EDS, at a cost of over $3 billion, and was also forced to take the lead in developing Manufacturing Automation Protocol (MAP) as a means to allow its 200 000 programmable machines to communicate. Despite this massive commitment to advanced technology, the most successful innovation in recent years has been GM's joint venture with Toyota at Fremont California, the NUMMI project. This old plant is consistently outperforming on productivity and quality indices, and represents a powerful advertisement for Japanese-type manufacturing practices.[9]

4.2   COMPUTER-INTEGRATED – OR JUST *POTENTIALLY* INTEGRATED?

The above dramatic examples have hit the headlines from time to time, but for most firms the experience is less radical, although still problematic. The typical situation is that projects take longer than expected, or fail to deliver to expectations. And even when such systems are finally installed, they are often used in ways which only exploit a small fraction of their full potential. If we return to our earlier model of substitution and integration, we can often find examples of integrated technology being used in substitution mode: for example, expensive flexible manufacturing systems being used as little more than dedicated plant; or powerful CAPM systems which could be used to provide an integrated information network being used to support discrete local activities such as stock control; or CAD/CAM systems used as little more than electronic drawing boards.

Substitution	Integration
CAD systems used as little more than electronic drawing boards	CAD/CAM used to integrate design and manufacture and to cut lead times
CAPM systems used to support local clerical activities	CAPM systems used as organization-wide information and support systems
FMS used to improve efficiency and local machine utilization figures, or to support manufacture of product in dedicated mode	FMS used to reduce lead times and improve overall organizational flexibility

Table 4.3   Integrated systems used in substitution mode

The problem is that the traditions of local efficiency improvement make it hard to change to a new way of thinking, one which emphasizes the broader goals of effectiveness across the organization as a whole. John Diebold, one of the pioneering writers on automation in the 1950s, foresaw this difficulty:[10]

> One of the impediments to re-thinking of products and processes has been that the traditional division of responsibilities has the effect of localising the areas in which re-thinking is done. Almost by definition, however, re-thinking must be done on an extremely broad basis – viewing the objectives of the entire organisation as a whole. It cannot be confined to the product design department. It must be an attitude, a state of mind permeating the entire organisation.

4.3   THE PRODUCTIVITY PARADOX

Although a large number of firms across many different industrial sectors have been able to implement AMT successfully, when we look at a macro-economic level this pattern of improvement does not seem to emerge as productivity growth. Indeed, according to the OECD, there has actually been a slowdown in productivity growth in most nations during recent years.[11] Yet this is precisely the time when we might have expected the powerful improvements resulting from IT to show through in terms of better use of inputs to production and resulting increased productivity.

This so-called 'productivity paradox' has concerned a number of economists for some time – and there is no doubt that such a complex problem is unlikely to have a single solution. But one answer to this conundrum might be that although extensive investment has been going on in AMT, most of it has been at the *substitution* innovation end of our spectrum. In other words, it has mainly been employed to do what has always been done a little better; and from this kind of investment we can only expect, at best, improvements in the *efficiency* of individual operations rather than improvements in overall manufacturing *effectiveness* right across the organization.

There is some evidence to support this contention. In a recent (1987) PSI survey an attempt was made to classify the kinds of IT being installed, the results suggesting that the vast majority – around 80 per cent – was of the substitution innovation kind, rather than more complex and integrated systems. As shown in figure 4.1, only 24 per cent of applications were even partially integrated and more highly integrated configurations approaching CIM were only present in 8 per cent of cases.

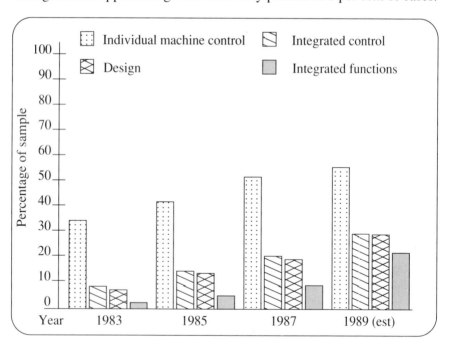

Figure 4.1   Levels of integration in UK manufacturing
Source: **Policy Studies Institute, 1988**

If we look at the type of equipment being installed (see figure 4.2) we can see that most of it is made up of relatively simple controllers for stand-alone applications: PLCs (standard controllers which can be tailored to particular – but relatively simple – control applications) are by far the most common application.

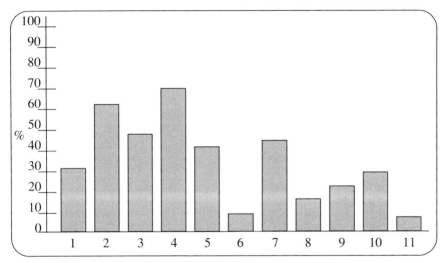

Figure 4.2   Diffusion of IT by type of application in the UK, 1987. 1, Design; 2, control of individual machines; 3, process control of individual items of plant; 4, machine or process control, or both; 5, automated handling (products/materials/components); 6, automated storage; 7, testing/quality control; 8, centalized machine control (groups of machines); 9, integrated process control (of several stages of process); 10, centralized machine control or integrated process control, or both; 11, centralized machine control or integrated process control, or both, plus design, testing and automated handling. *Source*: PSI, 1988

Other surveys bear a similar message. Their findings suggest very high rates of diffusion for technologies such as CAD and shop-floor production management systems. But closer inspection reveals that the majority of this growth has come at the low-cost, PC-based end of the market, where the systems can be used to support local activities within a particular function, rather than in more integrated configurations which might support the whole business. Figures for 1988 suggest that around 73 per cent of the installed base of systems in UK engineering, for example, is at this end of the market – although the trend is increasingly towards more complex and sophisticated systems.[12] Nor is this effect confined to the engineering industries: in the 1988 survey of process industry users, around 84 per cent of the installed base of computers was at the microcomputer end, with a value per installation of less than £12 500.[13]

Data from the US suggests that this pattern is not confined to the UK. In the latest US Census most applications already in place were in discrete rather than more integrated form. For example, NC/CNC machine tools were used by nearly half of the firms in the sample. However, two trends are clear from this study: that larger firms are already beginning to utilize integrated systems more extensively; and that investment intentions in the future will see a growth at this end of the application spectrum.[14]

In the A. T. Kearney studies mentioned earlier, most CIM systems throughout the world are still only integrated to about 50 per cent or less of their potential.

4.4   SUBSTITUTION INNOVATION AND ADOPTION THEORY

A partial explanation of this phenomenon can be found in innovation theory. Amongst other factors, adoption depends on the characteristics of the innovation being considered. Rogers[15] identifies five basic categories, and it will be useful to look at these briefly here:

1   Relative advantage – the degree to which the innovation is perceived to be better than that which it is to replace. The key here is in the word *perceived*: it is irrelevant whether or not the technology is actually capable of significant improvements, but the person making the adoption decision must believe this to be the case. It is fairly easy to accept technologies which make improvements to local efficiency because there are well known yardsticks against which to judge them. So if a technology saves energy or labour, or works faster, or produces to higher quality, the advantages are clear. But it is harder to 'prove' the advantages of 'strategic' investments which may require an 'act of faith' to justify them because their payback characteristics fall outside what is normally acceptable, and where many of the benefits that might emerge are not apparent prior to the investment. At the very least, it is likely that potential adopters will be more cautious about entering into such technologies – and the result will be a slow take-up of more integrated forms of AMT.

2   Complexity – the degree to which the innovation is understandable to the user. Again the distinction here is between innovations which substitute for what has already been done, the basic principles of which are therefore fairly clear – and those which open up radically new ways of doing things, often for the first time. The latter are clearly much more complex and this lack of understanding may act as a deterrent to adoption.

3   Observability – the degree to which the improvements can be seen or demonstrated. This 'seeing is believing' effect works against

more radical forms of innovation in two ways. First, it is much easier to demonstrate the benefits of a substitution innovation over what it replaces because the yardsticks are clearly defined. Second, the relatively slow take-up of the more integrated systems means that there are fewer such systems around for other potential users to look at. The demonstration effect is very strong in industrial innovation, and has been used in a number of government programmes to try to accelerate the take-up of new technology.

4   Trialability – the degree to which the potential user can try out the innovation before making a full commitment to it. This 'test drive' approach again works in favour of substitution innovation. As we have seen there has been considerable expansion of choice in IT products, with much of the growth coming at the low-cost entry-level systems, many of them based on PC technology. These provide a low-risk way for firms to enter a new technological field and to make mistakes and learn – without risking a great deal. By the same token many firms have made use of an 'islands of automation' philosophy, aiming to acquire advanced technological capability in piecemeal fashion. Such approaches have attractions in that both the cost and risk can be minimized, whereas investment in major integrated systems, although they may offer much greater strategic benefits, also represents high risk. The problem is that little opportunity exists for trying out integrated systems before commitment to them – it is very much a case of 'jumping in at the deep end'.

5   Compatibility – the degree to which the innovation fits with the context in which it has to operate. The key factor here is that it is possible to make substitution innovations work in the existing context because they do not require change in the way things are done Indeed, their development originated in an established context, with conventional assumptions about working patterns, levels of skill, organizational structures, and so on. By contrast, radical integrated innovations require major adaptation on the part of users, and this is a time- and resource-consuming process which can act as a barrier to adoption.

Therefore, one explanation for the 'productivity paradox' might be that although extensive investment is taking place in IT it is not the right kind of investment to have a major impact on total factor productivity growth. But even when that investment in integrated systems does take place it does not automatically mean that the full benefits will be obtained. As we saw in the preceding section, very often an integrated system will be used in a substitution mode, fulfilling only a fraction of its potential.

The problem with such investments lies in their complexity and in the degree to which organizations have to adapt and learn to use them. And the extent of adaptation required can be significant: not just learning new skills but learning completely new ways of organizing and managing manufacturing with these new tools. As a recent report from the OECD[16] put it:

> . . . neither the technical, nor the ecomomic, potential of major new technologies can be fully realised without concomitant, even anticipatory, social and institutional changes at all levels in society.

4.5   SECRETS OF SUCCESS

If some firms are able to exploit the benefits of AMT more successfully than others, it follows that they must be doing something different – especially when the technologies used are broadly similar. Two themes emerge repeatedly from research in this area. Success depends first on a clear link to *strategy*: if technology is purchased without a clear understanding of how it will support the business by facilitating competitive performance in manufacturing, then no matter how powerful or advanced it may be in theory, in practice its adoption will be little more than a high-risk gamble.

The second point is that success depends not just on adoption of a particular technology but also on the *implementation* of that technology in its organizational setting. The process of implementation is a long one and involves many non-technical dimensions (such as the management of organizational change). Unless firms are prepared to adapt organizationally then, once again, technological innovations which appear to offer powerful competitive advantages will not achieve their full potential performance, and may even *reduce* the effectiveness with which the business operates.

Several research studies confirm this. For example, Jaikumar's comparison of FMS in the US and Japan suggests that Japanese users of the same types of system were able to make them perform more productively *and* more flexibly than their US counterparts. What this highlights, according to Jaikumar, is that there are important differences in the way the systems are managed in the two countries.[17]

Similar findings are reported elsewhere. Indeed, as one commentator put it, in many ways the question of acquiring technology is no longer the issue, since it can be bought, licensed or even stolen from a variety of sources. What determines whether or not a firm gains any competitive advantage is how that technology is used.[18]

The management of technology is increasingly becoming an issue of strategic importance. Companies are now recognizing the competitive advantages which can follow from careful attention to product and process innovation, and from development of effective systems for managing such technological changes.[19] In the area of process innovation – that is, changes in the way in which products are made or services are delivered – there is growing recognition that a key success factor is simultaneous innovation in technology and in organization. To get the best out of new technologies requires 'a new way of thinking'. Traditional approaches to the organization and management of such technologies need to be reassessed and in many cases significantly altered.

This seems to be borne out in the experience of user firms which have successfully exploited AMT. For example, in our studies of FMS users, many report that the majority of the benefits came not from the equipment itself but rather from the changes in the organization and management of production which it forced them to make.[20] And in some cases, the process of carrying out a feasibility study led firms to make organizational changes as a precursor to new capital investment. Having done so, they found that they had already achieved their aims (of flexibility, inventory reduction, and so on) and did not need to introduce FMS technology.[21]

In essence, there are two kinds of organizational response to new technology. The first is a response to the kind of discrete substitution which we saw earlier, and basically reflects the familiar learning process which we associate with new equipment or with the manufacture of new product ranges. Thus it involves the acquisition of new skills and the development of familiarity in the use of the new equipment so that, over time, the full benefits become evident.

But the second reflects the augmentation effect, which only begins to emerge when some degree of integration takes place and where new ways of operating become possible. At this level there is not only the challenge of learning along familiar dimensions, such as skill acquisition, but a whole new set of issues raised by the need for organizational change. These would include:

- structural shifts (such as the changing role of functional differentiation between departments)

- work organization changes

- hierarchical changes (associated, for example, with the devolution of autonomy)

- strategic changes (to exploit the new opportunities which technology opens up)

- cultural changes – changes in 'the way we do things round here'

This also focuses attention on the key variable – the way in which change is introduced and is managed. The biggest challenge for management wishing to ensure the effective implementation of AMT lies in these areas.

4.6   ORGANIZATIONAL INNOVATION AS A SOURCE OF COMPETITIVE ADVANTAGE

Further support for the view that some degree of organizational adaptation is essential in making sure that AMT can deliver its full potential benefits can be seen in the growing number of examples which demonstrate significant improvements in the dimensions of competitiveness. For example, a major telecommunications equipment supplier was able to reduce inventory by over 70 per cent and to increase flexibility such that its original lead time of 14 weeks could be cut to a matter of hours. At the same time its quality record improved dramatically.

In another case, a firm in the engineering sector was able to cut its work-in-progress from 15 000 units down to 1000. Again, this was accompanied by improvements in quality and in flexibility, the firm becoming able to offer much better customer service including rapid response to changing needs.

Another firm in the shoe industry was able to cut batch sizes down from 1800 to 18 and still produce economically and to a higher quality standard than before. Along the way, its ability to respond flexibly and rapidly to the fashion market was massively increased, and it was also able to cut inventories and the factory space needed to hold them.

In a recent report, an IBM senior executive was quoted as saying that of all the company's $25 billion expenditure in factory automation in recent years, the most effective had been the least expensive, the projects involving continuous flow manufacturing (IBM's version of just-in-time).[22]

These appear to be impressive demonstrations of AMT's power to improve performance along many of the key dimensions of competitiveness for the 1990s – lead times, flexibility, inventory reduction, quality improvement, and so on. But it is significant that in each of these cases the benefits were obtained with *no investment in AMT* but, rather, by concentrating on organizational changes, primarily associated with the Japanese-derived ideas of total quality management and just-in-time production. These were not cost-free innovations – they incurred significant costs in the training area, for example – but they were more rapidly implemented, and at much lower cost, than investments in hardware would entail.

4.7   TECHNOLOGY AS A TOTAL SYSTEM

It is important to see such examples not as alternatives to AMT but rather as complementary to it. It should not surprise us that organizational change can contribute to improved flexibility and quality, any more than the fact that manufacturing systems based on advanced computers and machines also require organizational change to make them work. These findings simply remind us of the original and broad nature of 'technology', a dictionary definition of which is 'the useful arts of manufacture'.

Viewed in this broad fashion, technology is essentially a *system* involving both tools and the organization of the use of those tools. It is this point which helps us answer the question of what has to be managed in an AMT system: it is the *total* system, not just the physical components of computers (and their accompanying software). In those cases in which AMT has failed to work properly or to deliver its full benefits the problem lies in the relative neglect of change along the organizational dimension. As Perez points out, there is a growing mismatch of the technological and the organizational dimensions, and this threatens to retard the pace at which improvements in performance and competitiveness that might arise from the use of advanced technologies can take hold.[23]

This is particularly significant when we look at what might be termed 'revolutionary' technologies, those which transform the entire economy. We saw in earlier chapters that there is some justification for believing that we are witnessing the emergence of a new techno-economic paradigm, driven by the widespread availability of information and communication technologies. If this is the case, then the kinds of social and institutional adaptation required to ride this wave successfully are likely to be significant, challenging much of what is currently held to be 'best practice'. Put very simply, technological revolutions require similarly radical changes in organization.

This is a theme to which we will return throughout the book as we look at various examples of AMT. The strands are brought together in chapter 12, where we consider the implications for organization design which technology raises, but at this point it will be useful to outline the main areas in which such changes are taking place.

- Individual: new skills, multiple skills, flexible working and continuous training

- Work organization: multiple skills, team working, flexibility and responsible autonomy within cells

- Inter-group: functional integration, matrix and project teams, shared resources and de-specialization

- Control: flat, participative rather than procedural structures, local autonomy rather than rules, decentralization and networking

- Culture: organic and flexible, shared values/common purpose and commitment through shared ownership

This gives a broad indication of the direction and nature of changes required for effective implementation of AMT. The contrast with the present paradigm, with its emphasis on division of labour, tall hierarchies and emphasis on formal procedure and control, is clear. Arguably, such models will offer more flexibility and be better adapted to the challenge of dealing with uncertain environments in the future. But it is also important to stress that the detailed design of organizations for the new paradigm is only at an experimental stage, and there is no clear 'right' answer or model available for firms to use as a blueprint. Instead, there is a range of choice, running from the continuation of the old models, trying to make new technologies work within that context, right through to radical changes in organization design.

One of the most systematic studies of technological change and competitive performance has been the 'Future of the Automobile' project coordinated by researchers at the Massachussetts Institute of Technology. A key finding in their work has been a newly emerging model for manufacturing which owes much to Japanese experience but which is, to a greater or lesser extent, transferable. This concept of 'lean and fragile' manufacturing emphasizes a high degree of organizational flexibility and responsiveness in both product and process innovation, so that the firm is in a state of constant change and development, underpinned by principles of organizational learning and continuous improvement. Such factories use advanced technology where appropriate, but also employ a variety of organizational innovations (based loosely around the principles of 'just-in-time' and 'total quality management') to achieve their goals.[24]

In contrast to the 'lean and fragile' model, which achieves high performance on a variety of productivity and quality indicators, a second model is presented to describe the typical approach of many firms (and not just those in the car industry). This 'robust and buffered' model is essentially based upon a fixed view of how production should be organized and managed, and builds in a variety of slack resources to help to cope with unforeseen contingencies. It is relatively inflexible, slow to respond and is better suited to an environment that is characterized by low levels of uncertainty: above all, it does not contain the capacity for extensive learning, development and renewal. Innovation is seen as something embodied in advanced technology rather than a capacity of the organization as a whole, and such plants often have higher levels of automation than 'lean and fragile' plants: yet their performance is generally inferior, particularly in the important non-price factor areas of competitive performance.

Arguably, a key feature which will determine the degree to which firms can adapt successfully will be their capacity to learn from experimentation. AMT, especially in its more integrated forms, is not a deterministic technology with a single 'right' form of best practice. In many cases, it is not even a clearly defined physical configuration, but instead offers a range of choice in which the user plays a key role in further development. Given this malleability of both the physical technology and the organization, the key management skill for the future is likely to be that of a sculptor, shaping both to meet the needs of the business and the needs of the various stakeholders within it.

Notes

1   A. T. Kearney Consultants, *Computer-integrated Manufacturing: Competitive Advantage or Technological Dead End?* (London, 1989).
2   C. New and A. Myers, *Manufacturing Operations in the UK 1975–85* (British Institute of Management/Cranfield Institute of Technology, 1986).
3   G. Waterlow and J. Monniot, 'The state of the art in CAPM in the UK', *International Journal of Operations and Production Management*, 7 (1) (1986).
4   E. Arnold and P. Senker, *Designing the Future – the Skills Implications of Interactive CAD*, Occasional Paper 9, Engineering Industry Training Board, Watford, 1982.
5   B. Burnes, 'Integrating technology, integrating people', *Production Engineer*, September 1988.
6   J. Bentley, reported at a seminar on *Integrated Manufacturing* Birmingham, 10 May 1989.
7   *Sotsialisticheskaya Industriya*, 16 March 1988; reported in S. Glazev, 'Integrated automation in the context of structural changes in a modern economy', paper presented at *United Nations ECE Seminar on CIM*, Botevgrad, Bulgaria, 25–29 September 1989.
8   *Computer Weekly*, 23 February 1989.
9   J. Krafcik, 'Triumph of the lean production system', *Sloan Management Review*, Autumn 1988, pp. 41–51.
10 J. Diebold, *Beyond Automation: Managerial Problems of an Exploding Technology* (McGraw-Hill, New York, 1964), p.53.
11 'Total factor productivity growth in the OECD area slowed from an annual rate of about 3 per cent in the period from the mid-1960s to the early 1970s, to about $3/4$ per cent in the 1973–79 period, with further marginal decline in the 1979–85 period'; A. Englander and A. Mittelstadt, 'Total factor productivity: macroeconomic and structural aspects of the slowdown', *OECD Economic Studies*, Spring 1988.
12 *Engineering Computers*/Benchmark Research survey of computer applications May, 1989.
13 *Works Management*/Benchmark Research survey on automation in the process industries, Benchmark Research, October, 1988.
14 *Manufacturing Technology, 1988* (U.S. Department of Commerce, Bureau of the Census, Washington D.C., May, 1989).
15 E. Rogers, *Diffusion of Innovations* (Free Press, New York, 1984).
16 OECD, *New Technologies in the 1990s: A Socio-economic Strategy* (OECD, Paris, 1988).

17   R. Jaikumar, 'Post-industrial manufacturing', *Harvard Business Review*, November 1986 pp. 69–76.
18   K. Persson, 'Obstacles, strategies and mechanisms', contribution to Joint Meeting on New Manufacturing Technologies and Industrial Performance, OECD, Paris – November 1988.
19   For a discussion of this, see D. Teece, ' Capturing value from technological innovation', *Interfaces, 18*, (3) (1988), pp.46-61.
20   This theme recurs in several studies; for example, P. Dempsey , 'New corporate perspectives in FMS', in *Proceedings of Second International Conference on FMS*, ed. K. Rathmill (IFS Publications, Kempston, 1983).
21   Reported in J. Bessant and W. Haywood, *The Introduction of FMS as an Example of Computer-integrated Manufacturing*, Occasional Paper 1, Centre for Business Research, Brighton Business School, 1985.
22   S. Caulkin, 'More means less', *Computer Weekly*, 23 February 1989.
23   C. Perez, 'Microelectronics, long waves and world structural change', *World Development*, 13 (3) (1985), pp.441–63.
24   Krafcik, 'Triumph of the lean production system'.

5   Flexibility in Manufacturing Systems

5.1 INTRODUCTION

Thus far, we have discussed the wide applicability of IT, and its potential for a very broad range of users and applications. But, as we have also seen in the last chapter, applications to date have only really begun to scratch the surface. Most use of IT has been of the *substitution* innovation type, helping to make improvements (albeit often significant) in the *efficiency* of existing processes and operations. The possibilities which IT offers for more integrated applications remain relatively unexplored.

Part of the reason for this is that such radical use of technology demands considerable adaptation on the part of the organization – and the ability to recognize and manage the strategic challenge.

The next few chapters look at applications of IT which offer just such a contribution. We begin with what has arguably become the key issue for the 1990s – *flexibility*.

5.2 NEED PULL: WHY FLEXIBILITY?

At the heart of the strategic challenges confronting manufacturers in the 1990s is a need to become more flexible, more agile in their operations and more responsive in the ways in which they deal with customers. The market will no longer accept products in 'any colour you like as long as it's black'; instead they want high variety, extensive customization and frequent product innovation. Flexibility is also required to cope with fluctuations in market demand, which may vary widely in response to changing tastes, seasonal factors, advertising and many other variables. Inability to meet market demand and excess capacity are both problems which can damage business viability and so firms are looking to find ways of using their capacity more flexibly. This is particularly true as markets fragment, so that even if a firm is able to retain its overall share of a market, the volumes involved will be less, and so it will need to look for ways of producing more smaller-volume batches in high variety.

Customers also want reliable and fast delivery and rapid response – a movement towards an era of 'just-in-time' supply, only requiring delivery of something just in time for it to be used or consumed. With increasing competition and customer choice comes the need for manufacturers to become more responsive and provide a better level of service both before and after sales; and one important component is the ability to respond to sudden changes and urgent orders from customers.

Another area in which flexibility is increasingly required is in the routing and scheduling within the factory itself. In order to cope with fluctuations in demand and the frequent introduction of new products (with its implications for retooling, and so on), firms either need to build many different factories or find ways of making more effective use of their existing facilities. In many factories it is quite common for products to spend most of the time waiting for something to happen to them (queuing, waiting for inspection, in transit between operations, in finished goods stores, and so on) rather than actually having value added to them in some operation. So it follows that any improvements in the way in which work flows through the plant will increase the effective capacity and thus the ability to process a higher variety of work.

Also, within the factory there is the need to make more efficient use of capital invested in plant and equipment. In the days of mass markets and low-variety production it was possible to invest in dedicated plant, designed to carry out specific tasks and operations associated with a particular product, in highly efficient fashion. But with the move to shorter product life cycles, smaller production volumes and higher variety, this dedicated investment is no longer viable, and firms must look instead to ways of using the same plant to produce a number of different products and product generations.

We can briefly summarize the different types of flexibility which this environment is increasingly demanding of manufacturers, as follows:

- Product variety: more different models, styles, colours and so on

- Product customization: increasingly tailoring products to suit the requirements of particular customers

- Product innovation: frequent changes of model and introduction of new products to the marketplace

- Delivery flexibility: delivering on short lead times and in quantities to suit customer needs rather than manufacturing efficiency

- Demand flexibility: coping with seasonality, fashion and other types of demand variation – matching capacity to demand

- Routing flexibility: the ability to route manufacture of products in different ways, in order to cope with machine breakdowns or other interruptions

- Plant/equipment flexibility: the ability to use the same plant and equipment for different products – general rather than special-purpose plant and equipment

Such challenges used to pose problems only for small batch producers or those concerned with 'fashion' markets, such as clothing. But, as we saw in chapter 2, the trend now is for this challenge to confront even high-volume commodity producers, and for demands for more flexibility to emerge right across the manufacturing spectrum. Emphasis may vary (for example, small-batch producers may be more concerned with delivery flexibility, while mass producers may be more worried about product life cycle and product variety) but the direction of change is clearly towards much higher levels of flexibility in all manufacturing sectors. Indeed, a growing number of producers are taking the strategic option of positioning themselves as flexible producers of smaller volumes, with a high degree of specialization and customization.

For example, in the car industry, firms such as BMW are increasingly targeting their products at the high-value luxury car market, and offering customization to the point where it is possible for a customer to visit new facilities (such as the Regensburg plant) and see his or her car actually being produced to a personal specification. Plants are being redesigned accordingly, with extensive investment in monitoring and control systems to support such individual manufacture. Nissan Motor has recently set up a computer link between its development bases in Japan, Western Europe and the US to develop cars specifically tailored to customers' needs. Foreign bases have access to data held in the main technical centre at Atsugi, Japan. The advantage of this system is that a drawing which would usually take a matter of days to be mailed can now be sent electronically in seconds – and the result is an increase in customization of basic models to suit local needs and customer requirements.[1]

But what *is* flexibility?

One of the problems with a concept such as flexibility is that it means different things to different people. Is it about product variety? Or is it being able to respond rapidly on delivery? Is it about the regular introduction of new products to keep a 'fashion' market satisfied? Or is it about being able to move things through the factory in different ways to avoid breakdowns and bottlenecks? In fact, flexibility is a multidimensional attribute, something which different firms need in different ways. Gerwin suggests that the key to understanding flexibility is to see it as a response to *uncertainty* in the environment:[2]

Uncertainty about . . .	*Requires flexibility in . . .*
• what products/models the market will want at any time	product mix
• what length of product life the market will require	changeover to new products
• what particular attributes customers will want	modification and customization
• reliability and breakdown patterns of equipment	routing through plant
• volume and timing of demand	capacity
• quality and standards of incoming raw materials	material use
• delivery timing	sequencing through plant

There are many types of flexibility, some of which have short-term effects while others have longer-term implications. For example, in a study of 12 UK users, Lim[3] identified a range of flexibility types, commenting that certain flexibilities (such as product, volume and expansion flexibilities) have *strategic* implications, while others (such as process and product flexibilities) have *operational* implications. Krafcik suggests that we need to differentiate between complexity and flexibility, and he defines complexity as the product/part variation at a given moment in time, whereas flexibility is the ability to shift product/part mixes over time in dynamic fashion.[4] Tidd reports on recent work in Japan to try to specify and measure different types of flexibility so as to provide a tool for the evaluation of flexible manufacturing equipment, again emphasizing the differences between different types and the need to take a broader view of the concept than simply 'product variety'.[5]

In his studies of flexible manufacturing in Italy, Camagni takes this point a step further, suggesting that there are three levels, or stages, of flexibility and that manufacturers make choices about technological solutions on the basis of these.[6] The first concerns what he terms the 'substitution' stage:

> . . . when standalone machines and robots are directly substituting workers within existing production . . . the second stage may be called the 'production integration' one, where different machines are integrated into flexible systems, in the form of FMSs or cells, in order to perform successive processes. The innovative intervention concerns, beyond the technological system *per se*, the redesign of layout and of the form and characteristics of the pieces . . . in this stage fall those cases of integrated product and process

innovation which are typical of high tech industries and in general of those sectors characterised by fast innovation rates and short product life cycles . . . the third stage . . . may be called the 'strategic integration stage' . . . in this stage, the product nature and its market image are redefined according to the new opportunities offered by the flexible technology, and the entire firm organisation is deeply revised.

This model, which is essentially similar to our substitution/integration spectrum, suggests that flexibility is something which can have short- or long-term and tactical or strategic implications, depending on how it is interpreted within the context of company strategy. As we shall see later, firms can make use of flexible manufacturing technology to serve a number of ends, ranging from operational improvements – such as labour saving or increased utilization – right through to strategic targets such as improved customer service.[7]

Slack[8] characterises two different aspects of flexibility which he terms 'range' and 'response'. Range flexibility refers to the range of states which a production process can adopt; so that plant A is more flexible than plant B if it can make more different types of product, manufacture at different output levels or react better to different delivery demands. But changing over is not always easy, and is never completely cost-free. It implies time delays and resource costs which represent the 'friction' involved in moving from one state to another.

Response flexibility refers to the ease with which the changeover can be made, in terms of cost, time, organizational disruption and so on. For example, a plant may invest in an expensive flexible manufacturing facility, which gives it response flexibility since the investment is design to allow rapid changeover of products. But it is only designed to work within a distinct envelope or family of products, which effectively constrains the *range* flexibility. In terms of a wider product portfolio the investment may even turn out to have made the firm less flexible, because the overall range of flexibility is reduced.

In practice, response flexibility is essentially a short-term requirement since it refers to ways of directly improving the flexibility (service, variety and so on) associated with a particular product or market. But in the longer term range flexibility is important and has implications for investments in equipment, labour and production organization, since making the firm capable of a wider range of flexibility involves changing or adding to its manufacturing resource base.

For any system there will be several dimensions of flexibility, and each of these has its range and response flexibility components:

1 *Product flexibility* – the ability to introduce and produce novel products or services, or to modify existing ones. In this context,

range flexibility refers to the range of products or services which the company is able to produce, while response flexibility refers to the time necessary to develop or modify the product and processes to start new production.

2   *Mix flexibility* – the ability to change the mix of products or services being made. Here range refers to the range of products which the company can produce within a given time period, while response is concerned with the time necessary to adjust to the mix of products being made.

3   *Volume flexibility* – the ability to change the level of aggregate output. Here range refers to the absolute level of output which the company can achieve for a given mix, while response refers to the time taken to change the level of output.

4   *Delivery flexibility* – the ability to change planned delivery dates. Range is the extent to which delivery dates can be brought forward, and response is the time taken to reorganize the system so as to replan for the new date.

Meeting the challenge

In the face of this challenge for greater flexibility, firms have adopted a number of strategies. Some have attempted to move away from the specialization implicit in this kind of manufacturing and to move the basis of their competitiveness towards price, concentrating (or perhaps gambling) on being able to produce enough large volumes of a narrow product range to be able to exploit economies of scale and productivity improvements to the point at which they can sell more cheaply than their competitors. Such a strategy is very risky in a world of dramatically increasing competition, especially when many newly industrializing countries have significant labour or materials cost advantages, and in markets which are fragmenting and moving away from high volumes of demand.

A variant on this theme can, perhaps, be detected in patterns in the world automobile industry. Tidd suggests that, in the US, emphasis is being placed on focused manufacturing, going back to the basics of producing one product well. He cites Krafcik in support of this: 'a current trend seems to be the deproliferation of products and combinations on the part of US and European producers just as Japanese producers increase their model range and option availability in most world markets'. Data comparing product variants and product life cycles in the UK and Japan lends strong support to this view. The danger here is that emphasis on improving efficiency of manufacture by focusing on a narrow range, and using advanced technology to reduce costs, may be repeating the Model T problem – in the end the marketplace requires firms to try to resolve the 'productivity dilemma' and produce high variety with high efficiency.[9]

A second approach is to try to manage internally for reducing the need for flexibility in the manufacturing system, by reducing the product range or by making stock to cope with demand fluctuations. This kind of strategy carries serious cost penalties in terms of stock held 'just in case' of unexpected demand.

Another, more workable approach is to try to move towards some degree of modularity in product and process design, so that the flexibility in products delivered to customers is actually achieved at a relatively late stage in the manufacturing process, and much of the basic manufacturing is of a standard nature.

However, in general terms, firms are being forced to 'bite the bullet' and find ways of becoming more flexible, both in the short term, in trying to make their existing facilities more responsive, and in the longer term, in extending their range flexibility through appropriate investment and reorganization, as indicated in figure 5.1.

This is not always easy. A productive plant is, first and foremost, one set up to make the most efficient use of inputs to production, and so we might expect the highest productivity in plants dedicated to manufacturing one product in high volume.

Time is only one of the problems associated with trying to produce greater variety. As soon as the process becomes interrupted, other problems emerge:

- low machine utilization (due to set-up times for different batches)
- queuing problems at key 'bottleneck' operations through which all products must pass
- low machine utilization due to queuing upstream and waiting downstream of bottleneck operations
- high inventory levels of raw materials, work in progress and finished goods
- long production lead times
- poor production monitoring and control
- high overheads in indirect staff engaged in trying to monitor and expedite orders
- poor delivery performance
- poor quality

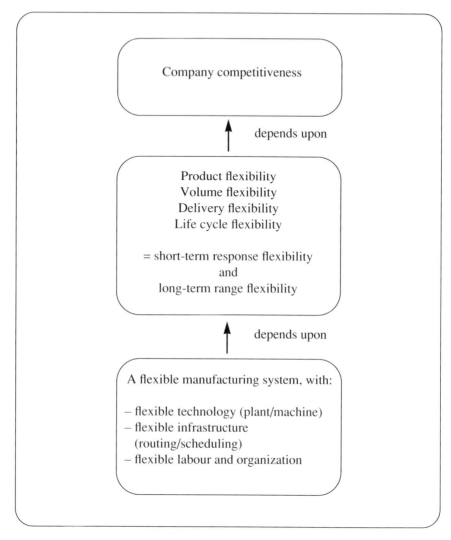

Figure 5.1   Flexibility and competitiveness

- inefficient use of space

- overloaded paperwork systems

These problems apply right across the manufacturing spectrum. For example:

- In a firm making soups, every time the product changes – say, from tomato to beef flavour – the mixing vessel and associated pipework needs to be cleaned out, the weighing equipment needs to be reset for new quantities, the pumps need to be adjusted for different dosage rates, the timers on the heaters need to be reset, and so on.

- In a firm making toys, the paint spraying operation needs to be cleaned out between different colours and reloaded with new paint. The equipment needs to be dried, the spray path reset or the paint sprayer instructed about the painting require-ments of the new toy, and so on.

- In a firm making pumps the machine tools need to be stopped, the tools changed, the machine tool feeds and speeds reset, the handling equipment altered, the fixtures realigned, and so on – all in order to facilitate changes in the size or model of pump being produced.

In essence, there is an implicit trade-off between efficiency and flexibility. Ideally, firms would like to produce large batches with minimal variety since this minimizes disruption. The best situation is a continuous-flow manufacturing operation, in which material flows through the system and there is no interruption or build-up of inventory. The pattern in many industries has been to try to approximate these conditions; for example, the mass production car sector makes use of special-purpose automation such as the transfer line to achieve this effect.

The problem is that, right across the spectrum, market pressures are forcing firms away from these solutions and towards those which offer ways of delivering higher levels of flexibility – not just in variety but also along other dimensions.

Abernathy[10] highlighted the problem of what he called 'the pro-ductivity dilemma', the apparent conflict between the desire to reduce costs and maximize productivity and the need to introduce new products. But evidence is now emerging that new technological opportunities and strategic pressures are beginning to force a resolution of this apparent conflict. For example, in his analysis of recent data on manufacturing strategies, de Meyer[11] suggests that the emergent trend into the 1990s is placing considerable emphasis on flexibility. Whereas in 1975–85 European and US manufacturers discovered that quality and cost efficiency were not incompatible, in 1985–95 events could prove that the traditional trade-off between flexibility and cost efficiency is similarly a thing of the past.

This opens up the question of what have been termed 'economies of scope'; which Tidd, paraphrasing Teece, defines as arising '...where the cost of joint production of several outputs is less than the cost of producing each output separately.'[12] If it is possible to achieve such economies, whether through the use of new technology, new organizational forms or whatever, then considerable opportunity is opened up across the manufacturing spectrum. In particular, the traditional bias in favour of large firms able to exploit economies of scale can be challenged by smaller firms producing specialized products targetted at particular market niches – a strategy which has become known as 'flexible specialization'.[13]

So, the need pull which has brought so much emphasis to bear on flexible manufacturing technology is made up of two basic components: the growing need for flexibility across the manufacturing spectrum; and the high cost of delivering this with conventional plant and equipment. In trying to explain why flexibility, of all the set of manufacturing system attributes, should be of such topical importance in considerations of company strategy, Slack suggests that '. . . the answer must rely on the unique position of flexibility to contribute to every measure of manufacturing performance, as judged by the customer'.[14]

5.3 TECHNOLOGY TO THE RESCUE: THE EMERGENCE OF FLEXIBLE MANUFACTURING SYSTEMS

In some industries the problem of flexible manufacturing has always been a headache, and a wide range of often ingenious solutions predates the emergence of IT by a considerable margin. For example, the use of Jacquard cards (punched cards which guided the movement of needles and hooks and springs) in the weaving industries at the turn of the nineteenth century meant that it was possible to produce different patterns with the same equipment simply by changing the sequence of punched cards. This made it possible to adapt relatively quickly to changing fashion markets (such as for floral silk patterns) and to make much better use of the looms. Similar developments were going on in the engineering industry which developed sophisticated arrangements of cams and gears to permit some degree of automation of operations.

In essence, there are two key ways in which AMT can contribute to increased flexibility; through *reprogrammability* and through *integration* of several discrete functions into a single system. The traditional approach to automatic control of industrial processes is to have some form of information about the desired activities held in physical form; for example, the punched holes of a Jacquard card or the cam of a machine. Such approaches allow a certain degree of automation, in the sense that machinery controlled in this way does not require direct input of information from an operator but can be pre-programmed. With the advent of electronics an increasing number of control systems began to utilize such pre-programmability; for example, in the sequencing of activities via relays instead of mechanical timers, and so on.

The problem with flexibility is that it requires a changeover in the setting of machinery and process equipment. Even with pre-programmed controls, changeover requires a complete replacement of the physical controls; new Jacquard cards, new cams or, in the case of early generation electronics, a new control circuit 'hard-wired' into place. IT offers major improvements because such changes to the control program can be achieved in *software*: the instructions for different activities can easily be changed and updated simply by writing a new program, rather than the more complex physical changes required previously. Further, the ability to control many functions through electronically operated actuators means that the resetting times can be dramatically reduced.

A second, important way of reducing the problem of flexibility is to integrate more functions into a single machine. In other words, by increasing the complexity of operations which one machine can perform we extend its range flexibility. Instead of moving between several different machines, each capable of a single operation, the process is integrated into a single machine or process stage. We can see this trend towards increasing flexibility through integration in the case of the development of metalworking machine tools.

The first machine tool for metalworking, the lathe, was invented by Maudsley in 1800, at the height of the Industrial Revolution. It involved a high degree of flexibility; for example, in the ability to change gears, change feed and speed rates, handle different-sized pieces of work and carry out various different types of turning. Its invention signalled the first moves away from the craftsman model in which a product was carefully and individually crafted to suit a customer, and towards more systematic production in a factory context.

So effective was Maudsley's design (it was, for example, used to cut the screws for astronomical apparatus) that it became the basis for the development of other machine tools for different tasks, which dominated the industrial scene for over a century. The next major stage in the engineering industry emerged on the other side of the Atlantic and was essentially a series of responses to the challenge of large-scale, high-volume manufacturing. Whereas the British tradition had largely been one of 'making', it was US industry which really laid down the basic ideas behind 'manufacturing'. This involved concepts such as that of functional layout – grouping machinery according to type of operation – and the notion of the 'mechanization of work'. As Jaikumar puts it:

> Whereas the English system saw in work the combination of skill in machinists and versatility in machines, the American system introduced to mechanisms the modern scientific principles of reductionism and reproduceability. It examined the processes involved in the manufacture of a product, broke them up into

sequences of simple operations, and mechanised the simple operations by constraining the motions of a cutting tool with jigs and fixtures. Verification of performance through the use of simple gauges insured reproduceability. Each operation could now be studied and optimised.[15]

This process involved the systematic integration of functions of individual machines and their operators into an increasingly complex manufacturing system, of which the Ford production lines represented a highly developed example. One part of this process was a gradual substitution of machines for men when and where it became possible to mechanize standard operations in the manufacturing sequence. However, there were limits to this substitution until the development in the 1950s of numerical control.

Control, and particularly concepts of feedback, had been extensively researched during the Second World War in order to control gunnery equipment. Numerical Control (NC) technology for machine tools evolved out of a US Air Force programme in the late 1940s, and essentially provided a way of describing the entire complex process of producing a part in terms of a mathematical expression. This code, when read by a suitable device, would be translated into a variety of control actions in the machine tool. The form this generally took was a paper tape with holes punched in it, which could be read and translated by a suitable reader. Later versions substituted paper tape with magnetic tape and later still, with solid state memory chips.

The key difference in this system is that of reprogrammability. In NC any change in the products to be produced no longer required extensive physical changes to the way in which the machine was set up, but only a change in the control program itself. Thus, for the first time, the flexibility of general-purpose machinery could be combined with the precision and accuracy of special-purpose equipment. In the process the human component, as an interface translating the design information into machine activities, could be replaced by some form of information processing device, such as a computer.

While the principles of NC were established in the 1950s, the technology did not diffuse widely until the 1970s, when low-cost programmable control, based around the microprocessor, became available. (Jacobsson and Edquist[16] suggest that the development of the microprocessor reduced the costs of the controller by about 50 per cent.) This opened up the development of the concept of Computer Numerical Control (CNC), and other functions, such as tool change or part manipulation, also became automated. Technical development at this time was characterized by a rapidly increasing integration of functions associated with the machine tool, leading to highly sophisticated, multipurpose machining centres, complete with a range of support functions such as tool change, head change, transport and manipulation, and so on – all under computer control.

This progressed to the idea of Direct (or distributed) Numerical Control (DNC) in which more than one tool (plus associated functions) could be grouped into a manufacturing cell under the overall control of a larger, supervisory computer which would be responsible for work scheduling, routing of products and parts, monitoring status, feeding in new programs and so on. This was essentially a process of bringing many different machines into a configuration in which they behaved as if they were a single, complex and highly integrated entity.

From this the step to a Flexible Manufacturing System (FMS) was a very short one. In general, a working definition of an FMS would be some combination of machine tools, handling systems, transport systems and ancillary equipment, the overall control and management of which is under some form of hierarchical computer control. As we will see shortly, this basic definition needs substantial revision the light of current developments, but it will serve for the moment,[17] and the process of convergence is illustrated in figure 5.2.

Integration on its own is, of course, not enough to offer greater flexibility. What is also needed is *reprogrammability*, the opportunity to change the ways in which that integrated configuration behaves without extensive intervention to reset it.

Industrial robots

Of course, this is only one application, but the same trends can be seen in others relating to different sectors. For example, in the assembly industries – whether metal, plastic or electronics – the same kinds of flexibility problems have been experienced, and have traditionally been dealt with by labour-intensive manufacturing processes. But the emergence of robotics as a reprogrammable IT-based manipulation technology, coupled with special-purpose machines and cellular manufacturing concepts, has opened up considerable possibilities in this area.

The term 'robot' was first coined by a Czechoslovak playwright, Carel Capek in his play 'Rosum's Universal Robots', in which it was used to refer to automatons capable of carrying out a range of human activities. Experiments aimed at developing such devices for industrial applications date back at least to the Second World War but, once again, it was not until the emergence of IT that suitable control systems began to appear to facilitate practical robotics. The first patent is generally accepted to have been registered by a British engineer called Kenwood, to cover an automatic manipulator, and this was soon followed by a US patent registered by Devol in 1961. His patent was taken up by Consolidated Diesel who later set up a subsidiary firm, Unimation, which played an important role in the development and early diffusion of the technology.[18]

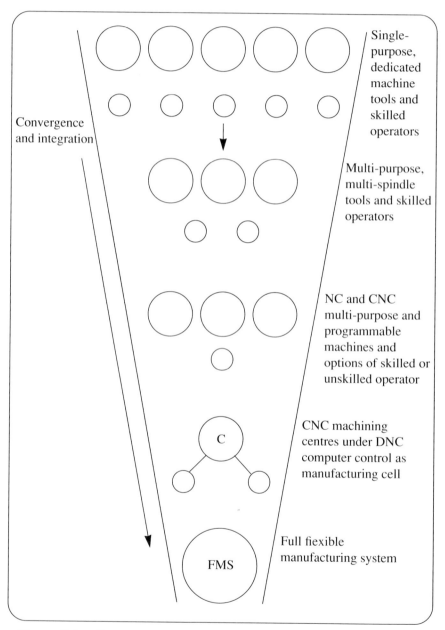

Figure 5.2   Convergence towards flexible manufacturing systems

Early robots were used in the vehicle plants of Ford and General Motors in the 1960s, but were mainly used for repetitive tasks such as diecasting. The Norwegian firm, Tralfa, developed the first tool handling robot for paint spraying in 1966, and a variety of firms followed this example. Welding applications emerged in the late 1960s but, in each case, the main applications were in high-volume series rather than for high-flexibility tasks in which reprogrammability would be important. ASEA in Sweden developed a robot that used electric rather than hydraulic drives in 1973, offering greater precision of control over movements, and the subsequent emergence of microprocessor control opened up possibilities in smaller batch work, especially in assembly areas. Unimation's PUMA (Programmable Universal Machine for Assembly) was originally developed for General Motors in 1978, but found widespread application in a variety of tasks.

Developments had also been under way in parallel in Japan, and in the late 1970s the SCARA robot emerged as the product of a major research programme. Like many pre-competitive research programmes, this involved 13 manufacturers, and the outcome was a general-purpose robot that was intended to be suitable for 80 per cent of all industrial assembly work, but at much lower cost than the more sophisticated PUMA type.

Despite the emergence of such general-purpose robots, much of the industry is still geared around task-specific robots, working on activities such as welding, spraying or long series assembly work. One of the problems is the lack of suitable sensors – vision systems, for example – which could make robots far more sensitive to variation and able to offer much greater flexibility.

Programmable controllers

In the process industries – such as food and drink manufacturing – the emergence of the low-cost programmable logic controller (PLC) and the possibility of networking this into hierarchies of computer control has opened up the field of flexible automation. Again, the two principles whereby flexibility is achieved are integration and reprogrammability.

The first PLC was developed in the 1960s by General Motors Hydromatic Division as a solid state control panel which could have its control functions changed without wiring alterations. By the mid-1970s, PLCs began to find application in many different sectors, and the supply side for such industrial automation took off. By the 1980s the original concept, that of a simple digital controller acting in isolation, had almost disappeared. The present generation of PLCs are powerful and support a very wide range of applications: digital and analogue signal processing, high-level languages for simpler programming, communication with other PLCs and within computer hierarchies, and beyond. Importantly, too, the product is available in standardized modules which can be built up into tailor-made systems to suit particular applications. Thus the same basic family of general-purpose PLCs can find application in industries as diverse as steelmaking, aerospace engineering and food processing. Not surprisingly, growth rates of applications of PLCs have been running at between 20 per cent and 30 per cent per year, the largest present-day markets being in food and drink manufacture and in pharmaceuticals and chemicals production.

In general, PLC applications can be used in a step-by-step pattern, moving up from simple materials handling, and applications such as packing or monitoring, where they operate in stand-alone mode, through sequential and interlocking control of processes right up to complex integrated machine control, energy management systems, and so on. Once again, the trend here is from substitution innovation to integrated strategic innovation.[19]

Applications of flexible manufacturing technologies

Each of these technologies can be used not only in their discrete form to enhance individual machine efficiency but also, through their reprogrammability, to improve the overall flexibility of operations. The higher the level of integration, the greater the potential flexibility improvement available.

In each of the examples given earlier, IT-based automation has facilitated greater flexibility. For example, pump manufacturers throughout the world have been able to use flexible manufacturing systems to reduce set-up times, and to achieve inventory and lead time savings resulting from a smoother and faster flow through the works.

A UK soup manufacturer was able to introduce a standard microprocessor-based PLC system capable of handling up to 200 different recipes and 30 different mix cycles: up to 24 of these recipes can be in production in the plant at any one time. In addition, the system automatically controls the clean-in-place and other process operations associated with changeover, and automatically monitors and adjusts the energy balance across the plant to ensure optimum use of heat.

In the case of the toy manufacturer, the use of a robot spraying system has combined automated cleaning and paint changeover sequencing with robot spraying. When new spray paths are required they can quickly be taught to the robot or programmed in directly.

In these and many other examples, the benefits of improved flexibility have come through the ease of reprogrammability offered by IT. This powerful edge can be further enhanced through the use of IT's capability for networking to facilitate integration, so that the bringing together of different functions is made possible – right up to the level of computer-integrated manufacturing (CIM). Along the way, we move from making improvements in local efficiency (increasing machine utilization, and so on) towards improving overall *effectiveness* through reduced lead times, inventory savings and better customer service.

Flexibility in textile manufacturing

A clear demonstration of the dramatic improvements which can be obtained can be seen in the case of J. and J. Cash, a medium-sized UK textile manufacturer. Cash's specialize in narrow fabric weaving and their best known product is quite literally a household name. Woven nametapes (with the name picked out in red thread on a white background), which are sewn into children's clothing to prevent loss, are generally known as 'Cash's tapes'.

For the company, this operation poses classic problems of flexibility management. Batch sizes are, by definition, very small (most orders are for the minimum quantity of 24 tapes) and the vast majority of orders come in a six-week period in summer, just prior to the start of the new school term. The traditional manufacturing process involved the use of Jacquard cards to control the weaving looms; 48 cards would be needed for each letter, so that a typical name of ten or more characters would require a string of around 500 cards. Every time a new name was woven, a new set of cards would have to be loaded on to the loom head – with obvious restrictions on its productive utilization. Although some economies could be made by keeping strings of cards for each letter in a library, the labour intensity of the card preparation area (where some 120 people were employed) was a major overhead cost.

The lead time for the production of nametapes using this system was between six and eight weeks, and many orders were delayed, especially during the crucial pre-school period, with the result that the company employed 27 people simply to handle customer complaints.

A further problem confronting the firm was the lack of suitable automated equipment with which to resolve this problem. Most textile equipment is made for the broadcloth sector, and relatively little technology is available for the specialist but small niche of narrow fabric weaving. Consequently, IT-based equipment, such as new looms, was not available, even if the firm had possessed the necessary financial resources to re-equip. Instead, they began the lengthy process of developing their own solutions, retro-fitting IT to their existing looms.[20]

The original process involved the receipt of orders from customers: these were then converted to Jacquard cards for weaving and the orders were scheduled for production. This stage, originally manual, lent itself to early computerization, and in the 1970s the process involved a large number of clerical staff typing in orders and names to a computer which then carried out production scheduling and generated the card preparation instructions.

The next stage of development involved the replacement of the Jacquard cards by paper tape, controlling a solenoid on top of the loom which was responsible for lowering and raising the threads. This inte-

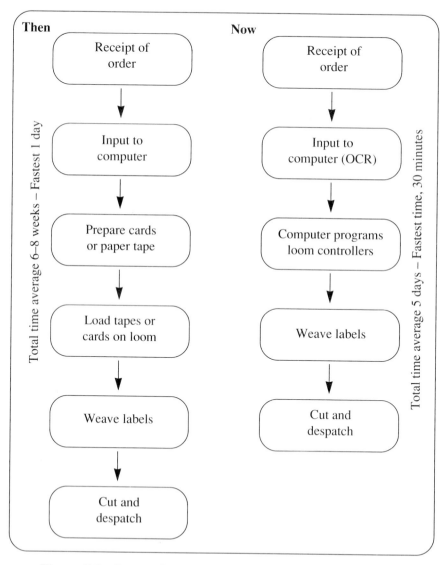

Figure 5.3   Increasing flexibility through convergence and reprogrammability

gration improved performance but was extremely expensive in terms of paper tape consumption (at one stage the company were the largest user of tape in the UK, using some 1500 km per year) and the tape readers suffered from wear and tear that led to a high incidence of errors. Further development led to a microprocessor control system which took direct control instructions from the mainframe computer, further reducing the time taken and cutting out the paper tape preparation stage.

The present system uses optical character recognition to input orders directly to the mainframe computer, which is still responsible for overall scheduling. From here weaving instructions are downloaded to the microprocessor controllers which supervise the weaving of nametapes. The tapes are then manually cut and packed, with the necessary production control and despatch information being provided by a printer connected to the mainframe.

The various stages in this long-term development are shown in figure 5.3. Flexibility improvements have been considerable, first through the reduction of the lead time from six to eight weeks to five days (or less, for urgent orders) to the point where all orders can be satisfied, even at the height of the season; and second, by extending the range of products available, through exploiting the greater capacity available on the looms (due to their more efficient use) and the fact that weaving control instructions are now held in software. A new style is now simply another program, and this has resulted in an expansion and development of the market away from name tapes in red and white, and in one size, and to a variety of sizes, colours, styles, and so on – all targeted at different market niches.

The key to this major improvement in flexible manufacturing capability (which has been extended to other areas in the factory) is in the use of reprogrammable technology, coupled with a gradual integration of functions into a single system.

5.4 DIFFUSION OF FLEXIBLE MANUFACTURING TECHNOLOGIES

The pervasive need for greater flexibility in the current manufacturing environment has ensured widespread and rapid diffusion of systems such as those described above. In this section we will look briefly at two examples of the more general trend, flexible manufacturing systems in metalworking and robotics.

Flexible manufacturing systems

It is generally accepted that the first FMS for metalworking was developed by Theo Williamson in 1962 for the Molins Company, a small batch manufacturer of machinery for the tobacco industry. His System 24 (on which the company still hold the patent for the name and design of a 'flexible manufacturing system') was able to machine blocks of aluminium into a variety of parts with short set-up and changeover times.[21] However, as with NC technology, it was not until the development of cheaper and more powerful control options that the technology of flexible manufacturing really came into its own.

One of the biggest problems in studying flexible manufacturing technology is the range of definitions which people use. Like Humpty Dumpty in Lewis Carroll's *Through the Looking Glass*, the label is often used to mean whatever people want it to – with the consequence that the label 'FMS' has been equally applied to a single, relatively simple computer-controlled machining centre and a 30-machine-tool factory! In the rest of this section we will use the following broad framework of definitions:

- flexible manufacturing system (FMS) – a combination of more CNC machine tools, under supervisory computer control via some form of DNC linkage

- flexible manufacturing cell (FMC) – a combination of two or more CNC machine tools but not under DNC-linked control

- flexible manufacturing unit (FMU) – a single, multifunction CNC machine tool

- flexible transfer line (FTL) – a multimachine layout including several CNC machine tools and other specialist pieces of equipment all under supervisory computer control, used in high-volume industries such as vehicle manufacture as an alternative to a dedicated transfer line

From the early 1980s onwards, diffusion of FMS has been rapid, although the number of systems worldwide is still relatively small. Estimates suggest that there are between 700 and 800 FMS installations in the world (using a cut-off definition of at least two CNC machine tools plus automated handling and centralized computer control).[22] On this basis, annual growth rates have been of the order of 30 per cent which, if continued to the year 2000, will imply several thousand systems worldwide. Even allowing for saturation effects this suggests a population of up to 3500 installations, and this does not allow for the emergence of new areas of application beyond metalworking.

The extent of penetration of this technology by country is also significant, with all major industrialized countries having some share in investment.[23] Emphasis on application is still for high product variety and small batch manufacture – as shown in figure 5.4 – but it should also be noted that there are applications in which flexible manufacturing is appropriate for large batch sizes, as in the vehicle industry, for example.

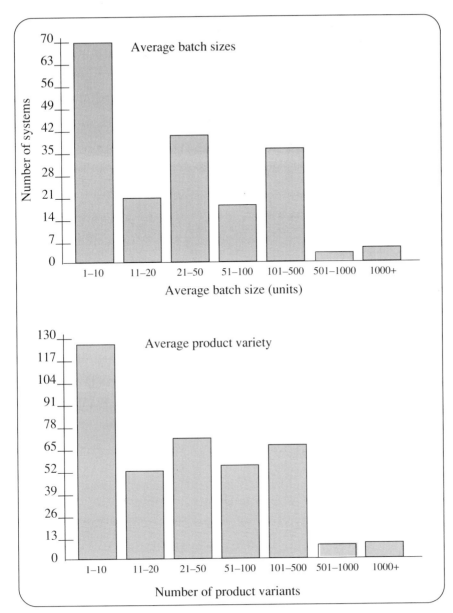

Figure 5.4 Application of FMS. *Source:* **IIASA, 1989**

There is also evidence that an increasingly wide size range of firms are taking up the technology, especially in its more compact form, as flexible manufacturing cells or units. Whereas FMS was born in the high-value and high-complexity businesses of aerospace, the original definition of FMS has undergone considerable revision. There is now a proliferation of choices across the manufacturing spectrum, with a distinct split into two groups, expensive/complex and simple/cheap/compact systems. This can be seen in the breakdown, both in terms of the number of machine

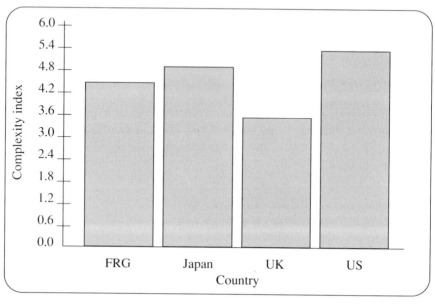

Figure 5.5   Breakdown of FMS by complexity

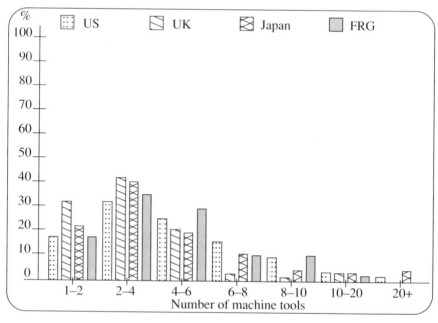

Figure 5.6   Breakdown of FMS by number of machines

tools (figure 5.5) in the systems and, more accurately, by using a specially developed measure of technical complexity (figure 5.6).

A typical case of a large system user might be that of an engine manufacturer, producing a range of five to eight cylinder heads in volumes of around 20 000 a year. Traditionally these would be made on three dedicated lines, which are replaced by a large FMS with a theoretical potential of considerably greater flexibility; for example, 100 different heads. Thus the system has long-term potential for adapting to market changes, although day-to-day operation will suggest only limited flexibility within an eight-head envelope. The greatest use of flexibility is for long-term model changes; for example, annual model change or introduction of new products: most benefits are likely to arise from operational improvements, especially in the area of inventory saving (around 70 per cent of WIP) and lead time reduction in this area (cut by 70 per cent) – although not necessarily across the business as a whole.

By contrast, a small system user would probably be a small- to medium-sized enterprise, often in the subcontracting type of business. In many cases the main justification will be for capacity improvement rather than for flexibility, since the traditional mode of operation is already highly flexible. Again, main benefits will be operational; savings in WIP, space and lead times.

Typical characteristics of these two types of systems are indicated in table 5.1.[24]

Compact	Complex
2–4 Machine tools	15–30 Machine tools
Conveyor transport	AGVs
Possibly automated store and retrieval	Automated storage and retrieval
PLC controller linked via LAN to local PLCs	Mainframe (VAX or equivalent), back-up computer as well
Two robots	Multiple robots

$3 million, of which:
  Machine tools = 50–55%
  Handling/transport = 15–20%
  Control and communications = 20–25%
  Planning and training = 10%
	$10–15 million, of which:
	  35–40%
	  15%
	  25–30%
	  15–20%

Relatively limited expansion	Open expansion potential

Table 5.1   Characteristics of different system types

The trend is clearly towards further growth in standard systems which, at least in the short term, will emphasize the diffusion of compact cells. As Ranta and Tchijov comment:

> . . . the higher number of CNC machines combined with a large part family usually results in such a complexity of systems co-ordination (for example, routing, scheduling, tool management) that high software and planning costs cannot be avoided. The only way to get the relative costs down is a modular systems structure and standardised software modules. The benefits of standard systems are already clearly visible in compact systems.[25]

This trend can be seen in figure 5.7, which indicates the distribution of FMS configuration in France and the Federal Republic of Germany, and in figure 5.8, on the increasing importance of small systems, based on data from France.

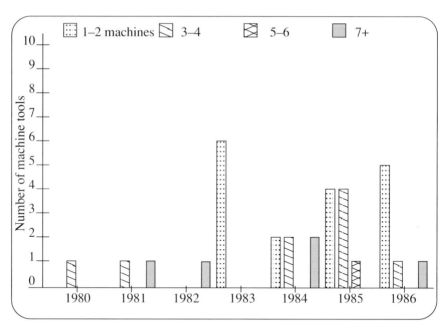

Figure 5.7   Distribution of FMS by type in France

There is a broadening-out across the manufacturing spectrum of options for flexible manufacturing so that, for example, relatively high volume users can exploit the technology of 'flexible transfer lines' (FTL), while small batch and prototype producers can also make effect-ive use of flexible manufacturing cells and units. This again reflects the

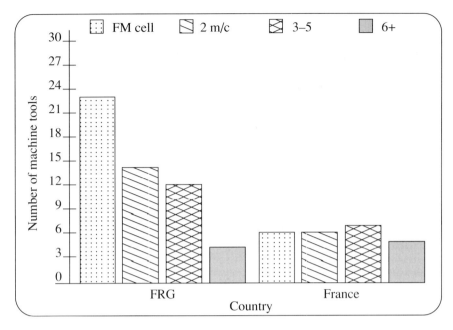

Figure 5.8 Distribution by FMS configuration, France and Germany

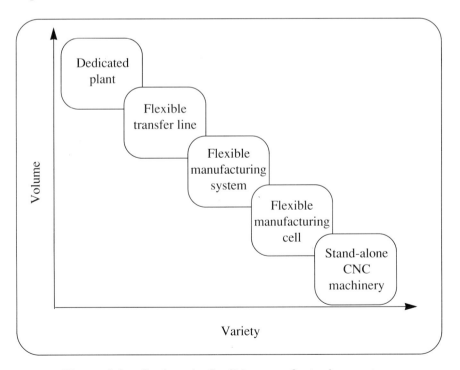

Figure 5.9 Options in flexible manufacturing systems

point that different users require different kinds of flexibility. For example, one supplier[26] suggests that '. . . with the latest 3 axis machines, head in-dexers and improved loading and feeding systems, flexible transfer lines can handle 20 parts/hour and above, leaving 'traditional' FMS – random processing of random parts – operating in the area of 0.5 to 10 parts/hour'.

Their motives for doing so also vary; from acheiving better life across multiple product life cycles in the car industry, to being able to manufacture very small batches economically in the aerospace sector. This broadening-out of options across the volume/variety product spectrum is illustrated in figure 5.9.

This proliferation is beginning to affect sectors beyond metalworking; such as woodworking, plastics and clothing. But the main concentration remains in metalworking. For example, current estimates suggest that around 80 per cent of systems are used in metalworking and, of these, 80 per cent are in cutting, 15 per cent are in sheet metal work and only 5 per cent are in flexible assembly systems.[27]

Industrial robotics and flexible assembly

Like FMS, robotics has begun to diffuse widely, although many experts still consider that present levels of application are low, being held back by the lack of suitably sophisticated and intelligent technology. This is particularly true for the assembly area, where lack of suitable sensors – for vision, touch, and so on – means that robots have so far only found limited application. An indication of the diffusion of the technology is given in figure 5.10; if the actual 'head count' is used, then Japan emerges as the dominant user, with close to 200 000 robots in use. The breakdown amongst major users is given in table 5.2. However, taking

Country	Number of robots
Japan	176 000
US	32 600
Federal Republic of Germany	17 700
Italy	8 300
France	8 026
UK	5 034
Sweden	3 042

Table 5.2   Main users of robots, 1988[27]

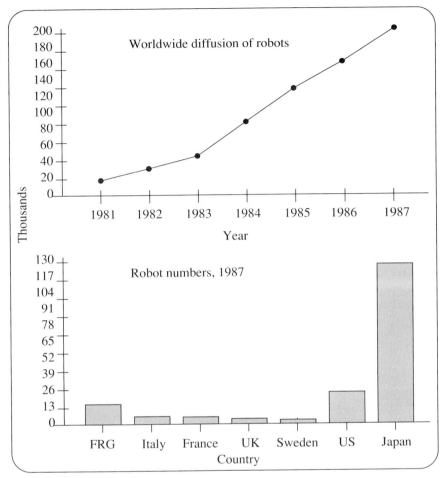

Figure 5.10   Diffusion of robot technology. *Source:* **Tidd, 1989 (compiled from various sources)**

the overall density of robot use (by dividing the num-ber of robots by the number of employees) makes Sweden the world's most intensive user of robots.

The distribution of robots by application is given in figures 5.11 and 5.12, the density figures providing a sharp picture of the concentration in the automobile industry. Finally, an indication of the size distribution of user firms, again based on density of application, is given in figure 5.13.

From the above we can discern a number of trends in the use of robotics technology. The first is the increasing diffusion across sectors, firm sizes and applications, with assembly automation set to become a much larger user in the medium-term future. Indeed, since the development of the SCARA robots in Japan for general-purpose assembly tasks, this area has experienced the most rapid growth in most industrialized countries.

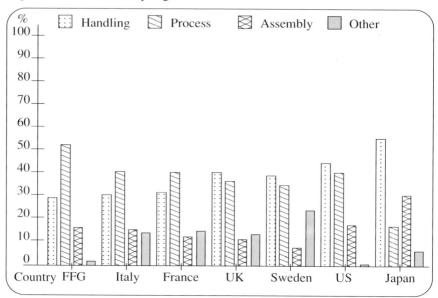

Figure 5.11   Distribution of robots: application by sector.
Source: **Tidd, 1989 (compiled from various sources)**

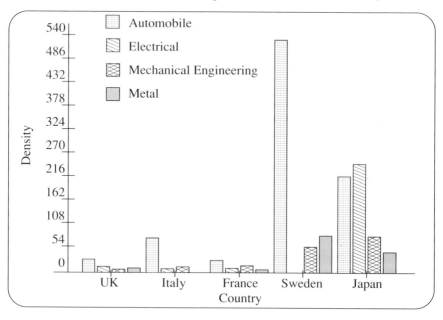

Figure 5.12   Distribution of robots: application by type

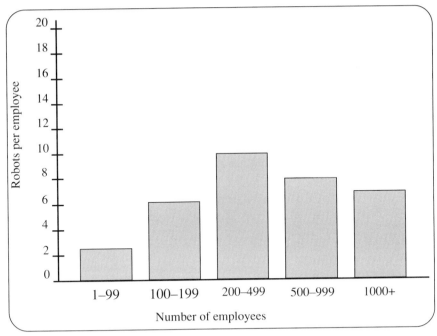

Figure 5.13   Distribution of robots by firm size

Amongst key technological developments currently being explored are:

● direct drive robots, in which motors are directly connected to arm joints rather than via gears and transmissions

● advanced sensors, especially in image processing for vision systems

● improved, more sensitive grippers

● artificial intelligence based programming so that robots can 'learn' from their own mistakes

Work has been going on in all of these areas, and several expensive and dedicated applications already exist. The real challenge is to create a 'second generation' of general-purpose robots with these characteristics.

Programmable logic controllers

PLCs have emerged as a relatively low-cost, general-purpose controller, and consequently have diffused widely. As we saw in the previous chapter, applications vary enormously and cover all sectors of industry. Some indication of the take-up can be gained from surveys which suggest that in the UK in 1987 around 50 per cent of firms had some form of PLC, and 30 per cent had CNC or related machine controllers.[29] In the US (1988 figures) the level of CNC-type control was around 40 per cent of firms, by far the most common form of flexible automation.[30] In the preceding chapter we saw how a single type of microprocessor could cover a wide range of food industry applications, and similar examples exist across the industrial spectrum.

5.5   EXPERIENCE WITH FLEXIBLE AUTOMATION

Motives for introducing some form of flexible manufacturing system are typically associated with trying to avoid the penalties – costs, time, excess inventory, and so on – which result from making things in smaller batches with higher variety. The extent to which firms have been able to achieve these kinds of advantages can best be judged by reference to some survey data and case examples. Data from the IIASA survey suggests that lead times, set-up times and work-in-progress inventory levels have all been reduced by over 50 per cent in the majority of cases of FMS implementation, and this experience is borne out in other research. For example, in studies of FMS in the UK and in Sweden, Bessant and Haywood report the benefits listed in table 5.3.[31]

Size of firm (employees)	Lead time (%)	WIP (%)	Utilization (%)
1–500	– 66	– 66	+ 45
501–1000	–76	– 63	+ 50
1000 +	– 86	– 70	+ 55
UK average (50 firms)	– 74	– 60	+ 54
Swedish average (20 fiirms)	– 69	– 60	+ 64

Table 5.3   Benefits of FMS use in Sweden and the UK

In another report on FMS experience in Finland,[32] based on a detailed study of 12 cases, the benefits reported for these systems were:

- reduced inventory – 70 per cent (mainly WIP and finished goods)
- reduction in number of machines needed – 66 per cent (but overall slight increase in capital cost of equipment)
- utilisation increase (based on 24 hours) – 72 per cent
- productivity increase – 170 per cent
- reduction of direct labour – 68 per cent
- size of investment – $2 million
- payback – 2.8 years

Another study in West Germany looked at 60 firms using a total of 95 FMS cells.[33] The benefits reported are indicated in table 5.4.

Factor	Percentage of firms achieving
Reduced lead time	57
Improved quality	36
Unmanned operation	27
Labour saving	24
Increased product variety	30
Reduction of downtime	24
Parallel set-ups	57

Table 5.4   Benefits of FMS use in the Federal Republic of Germany

In a survey of 31 FMS users in the UK, the main benefits which firms reported were the following:[34]

1   *Lead time and throughput (factory door-to-door) time reduction.* Here the general experience is that substantial savings can be made in throughput (factory door-to-door) time. Savings of between 60 per cent and 70 per cent were reported by 20 firms (65 per cent) in the sample, while two others also reported savings but could not quantify them. Reductions in the lead time to the customer were also quantitatively reported by 13 firms (42 per cent), again with two firms commenting on the importance of this factor without giving quantitative estimates. The lower figures can partly be explained by the fact that cutting the time to produce complex parts within the plant through reducing the set-up time and so on, does not of itself guarantee that the lead time to customers will be cut, since this time is dependent on several other elements which may not be affected by the FMS investment. Mention should also be made of two other cases in which lead time was crucial to the success of the business, and where the role of FMS had been to help the firms *maintain* their tight lead times despite increasing product complexity and more detailed customer specifications.

2   *Inventory savings (especially of work-in-progress).* Another expected consequence of reducing set-up times and integrating process elements is that there is smoother flow of material through the factory with less queueing and build-up of material waiting for machining. For the firms which provided a percentage estimate

the average saving was over 70 per cent, and in other cases figures ranging from £0.5 million to £2 million were saved. Overall, 15 firms (48 per cent) reported improvements, although in some cases these were slight; and in several the point was made that savings should be attributed to other innovations occurring simultaneously, such as just-in-time manufacturing.

3   *Increased utilization.* One major problem in batch manufacturing is the relatively low level of utilization of equipment, since so much time is spent waiting for products to be put on machines which are stopped for resetting, maintenance, and so on. In 12 cases (39 per cent) improvements were reported, and the extent of improvement over previous methods ranged from 200 per cent to 400 per cent. For others in the sample involving new plant there was no direct comparison which could be drawn with previous methods but, as indicated in the preceding discussion of current levels of usage, most FMS equipment is now operating at a high level of utilization.

4   *Reduced set-up times.* Closely linked to improved utilization is the reduction in set-up time between different batches. Twelve firms (39 per cent) reported improvements over previous methods of between 50 per cent and 90 per cent.

5   *Reduced number of machines or operations.* Part of the reduction in set-up times and in throughput times is derived from the physical integration of operations into fewer, more complex machines. In this sample ten firms (32 per cent) provided estimates of the extent of reduction of number of machines with the shift to FMS, ranging from 45 per cent to 90 per cent.

6   *Increased quality.* As with earlier examples, quality improvements (while sought from FMS investments) cannot be wholly attributed to that technology, since in many firms there is considerable effort being spent to improve quality performance; for example, through total quality programmes. However, 20 firms (64 per cent) reported improved quality as an outcome of their investments, with several others which involved new plant also commenting on the high levels which they were able to achieve. Quantitative estimates of this improvement ranged from 20 per cent to 90 per cent in one case.

In addition to the above benefits, a number of others were reported, often on a firm-specific basis. These included:

- space savings
- reduced dependence on subcontractors
- skill-saving

- increased responsiveness to customers (speed and quality of service)
- facilitated more rapid product innovation cycles
- improved prototyping capability

One indication of the benefits obtained from robot use comes from the 1985 UK survey, which is summarized in figure 5.14.

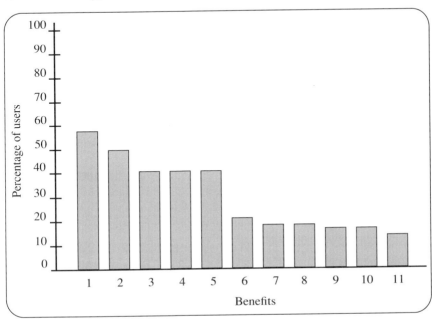

Figure 5.14   Benefits from robot use in the UK, 1985.

1, Improved quality, more consistent products;
2, lower labour costs;
3, greater volume of output;
4, improved working conditions/safety;
5, increased technical expertise;
6, better management control;

7, greater reliability, less downtime;
8, better labour relations;
9, greater flexibility for product changes;
10, lower material costs, less waste;
11, less capital tied up in WIP. Source: PSI, 1985

5.6   STRATEGIC OR TACTICAL INNOVATION?

In all of the above applications we see rapid diffusion and apparent benefits. But it is worthwhile focusing a little more closely on the kinds of benefit which have been achieved.

We saw earlier that flexibility could be categorized into short-term and long-term, with the former referring primarily to a general *operational* flexibility and the latter covering a much broader *strategic* flexibility. A variety of studies suggest that the bulk of FMS adoption is for operational flexibility motives; or even, in many cases, for traditional cost-saving motives which do not feature a very high increase in flexibility. And even where strategic flexibility is sought, it is often not achieved in practice.

One reason why 'traditional' motives dominate is that these are easier both to realize and to measure at the investment appraisal stage and beyond. For example, Tombak and de Meyer concluded that European and US firms do not buy FMS to change their product designs more frequently, but rather to standardize their product lines, using FMS to cover variability in inputs.[35]

Boer et al, report on a study of seven FMS installations in the UK, Belgium and the Netherlands: '. . . whatever the prime motives are, most FMS adopters are primarily aiming at operational advantages in order to pay back the investment. Using FMS to obtain market advantages is considered later' (five out of the seven did this). But '. . . only 1 of the 7 succeeded in achieving their operational goals in the time set'. The reasons for this included technical problems, changes in the marketplace (leading to under- or overcapacity) and insufficient organizational adaptation (with the FMS managed as a local innovation in one area rather than as an organization-wide integrating technology.[36]

In her studies of ten US FMS installations, Graham points out that getting operational benefits was much easier than obtaining strategic benefits.[37] Margerier's studies of FMS in France concludes, on the basis of 19 systems that, '. . . the findings of the survey leads us to believe that it is the reduction of unit costs which above all determines the choice of this type of equipment, rather than a strategy aimed at product diversity.'[38] Even in Japan, the motives cited in recent surveys suggest an emphasis on factors such as cost reduction; although this reflects in part the rising value of the yen in recent years, and the need to preserve an element of price competitiveness in world markets.[39]

In research into 31 cases of FMS installations (carried out by the Centre for Business Research and supported under a government scheme to promote adoption of this technology), a distinction was drawn, in the case argued for capital allocation, between the *motives* which firms cited and the actual *justification* for making these investments.[40] Market and competition-led motives underpinned most investment decisons. Typical responses here included the need to respond to increasing competitor threat by upgrading performance on several non-price dimensions as well as on prices themselves; for example, by offering higher quality, shorter delivery times, greater customization and product variety, more frequent product innovation, and so on. Prices depend in part on the efficient use of resources, and another group of motives were geared around the cost-saving aspects of FMS, such as the ability to minimize inventories.

Motives were not only driven by the need to defend or recover market share; a second group were concerned with opening up new opportunities, either by offering new products and/or a better service in existing markets or by opening up new markets with new products. Beyond these general market/competition motives a variety of specific motives were expressed, including:

- increasing quality
- the ability to handle volume fluctuations or produce in smaller batches
- offering greater product variety or customization
- reducing the lead time on manufacture and delivery
- improved customer service (for example, by involving the customer in the actual design or specification process through the use of integrated CAD)
- bringing in house activities that were previously subcontracted
- coping with skill shortages
- handling high-complexity tasks in unmanned fashion, with better accuracy and skill-saving
- greater in-house flexibility in routing, avoiding bottlenecks
- upgrading of equipment as part of a long-term plan
- 'because we ought to', investment for purposes of company prestige, and so on.

In several cases the motive of experimentation was also explicit: firms recognized the need to explore the technology.

Although the motives for investment covered a wide range of issues (some general, some firm-specific), the pattern with actual *justification* for the investment was much more focused. Here the majority of firms reported justification for investments based on either direct cost saving – in most cases of labour – or reductions in working capital tied up in inventory (primarily WIP). Other categories included savings in sub-contracting costs, shortened lead times (and hence better cash flow) and increased output. In addition, for several firms the FMS was tied to a new product (often accompanied by a major new capital investment programme of which the FMS was only a part) and, as such, was justified on the basis of expected sales.

Most firms attempted to justify their investments using standard appraisal techniques, but many confessed to serious reservations about this approach. This process was often seen as an exercise to obtain a rough idea of the costs and savings in support of what was seen to be a longer-term strategic investment.[41] A consequence of this was the need to relate investment to some tangible and quantifiable form of saving – such as in labour or inventory. The majority of projects were justified against savings in direct labour, which is significant since labour-saving did not appear as a key motive in earlier discussion. Arguably, firms are looking at FMS as a means of achieving strategic goals such as greater flexibility and non-price competitiveness, but are being forced by accounting requirements to justify these investments on the basis of much shorter-term criteria.[42]

It's not what you've got, it's how you use it....

All of this tends to suggest that we are still at the early, substitution stage of the use of flexible automation technology. The more strategic uses are still to come; but getting there poses problems because of the increasing emphasis which achieving full flexibility places on the way in which the equipment is used, rather than the physical nature and configuration of machines and control systems.

For example, Margirier[43] was able, for a small sample of seven cases, to calculate the potential flexibility of the FMSs and compare this with the flexibility actually achieved. He concludes: '. . . the effective flexibility of systems is not closely linked to their potentialities but rather depends heavily on the way the equipment is used'.

Factor	US	Japan
Number of different parts produced per system	10	93
Parts produced per day	88	120
Number of new parts introduced per year	1	22
Utilization rate (based on two shifts)	52%	84%
Average metal cutting time (hours per day)	8.3	20.2
Source: Jaikumar, 1986		

Table 5.5   Comparison of US and Japanese FMS installations

Jaikumar's detailed comparison of US and Japanese systems[44] bears this out strongly. In this study of systems which were broadly similar in terms of their physical characteristics, he found significant differences in the degree to which the Japanese were able to exploit both flexibility and productivity advantages when compared with the US firms. The main points are summed up in table 5.5.

This suggests that Japanese firms are able to manage their systems not only to be more flexible but also to be more productive than their US counterparts; that is, they appear to have gone much further along the road towards resolving the 'productivity/flexibility' trade-off identified by Abernathy and others. In his analysis Jaikumar attributes the differences primarily to differences in the way in which the Japanese systems are used and managed.

This experience is not confined to flexible manufacturing systems. Tidd, in his detailed comparisons of robot use in the UK and Japan, reached similar conclusions. As he points out, differences are emerging in the way in which robotics is used, with evidence that Japanese use is based on simpler, less technically sophisticated applications than in the UK. Despite this:

> Installations in Japan appear to be more flexible than those in the UK: in Japan more product variants are assembled by each robot system, and the product life cycles are considerably shorter. In the UK there is a strong relationship between the complexity of the assembly task and the sophistication of the robotics technology used. However, it appears that no direct correlation between the level of technology employed and manufacturing flexibility exists. This suggests that managerial and organisational factors are at least equally significant. For example, in Japan users concentrate on product and component design before implementing robotics, in the UK users attempt to 'automate out' deficiencies in such areas.

The difference is indicated in figures 5.15 and 5.16, which highlight the higher product variety and shorter product life cycles handled by robot systems in Japan.

Clearly, we should look closely at the ways in which the systems are managed and organized for flexibility. It can be argued that this is related to the degree to which flexibility is seen as a property of the total system, and not something which resides only in the hardware and software involved. The introduction of FMS, or robotics, or other forms of flexible automation is essentially a *configurational* exercise, involving simultaneous adaptation and development of a number of elements – machines, software, people, work organization, and so on. This places

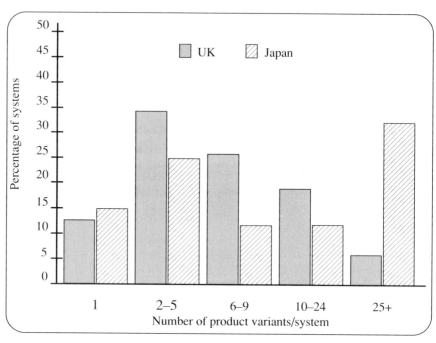

Figure 5.15   Product variety in flexible assembly

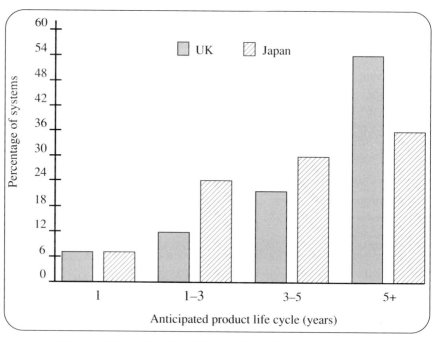

Figure 5.16   Product life cycle in flexible assembly

emphasis on the whole process of implementation, not just on the decision to adopt the technology, and it is to implementation issues that we now turn.

5.7   IMPLEMENTATION ISSUES

The importance of treating FMS, robots, and so on as configurational technologies has been highlighted by a number of writers.[45] Contextual factors, such as the layout of production, the choice of products to be made, the linkages between different support functions, and the skills and work organization of those operating the system, all have a bearing on how well such systems perform. Indeed, a point which emerges from a number of studies is that the benefits of FMS investment often come more from the organizational changes which it catalyses than from the narrow set of physical equipment which is installed. It forces a 'new way of thinking' on the firm, and it is this change in approach which is critical in obtaining the full benefits.[46]

In particular, this suggests that we need to look beyond the physical hardware and software options for sources of flexibility in manufacturing systems.

Automatic or computer-aided flexibility?

As we saw earlier, many flexible systems are justified on the basis of labour-saving, despite the fact that the prime *motives* for adoption may involve longer-term strategic goals. The inappropriateness of labour-saving as a basis for cost justification is highlighted when we consider that labour costs are often of decreasing importance, not least because in many engineering firms direct labour is now responsible for only a small element of total manufacturing costs – in many cases only 10–15 per cent – whereas materials costs can represent 40 per cent or more. (For this reason an increasing number of FMS installations are justified on the basis of inventory savings.) Ebel reports research undertaken in the Federal Republic of Germany which suggests that only 1 per cent of the workforce is working with FMS/CNC technology – and, further, that the total displacement of labour due to the introduction of FMS/CNC is estimated at around 2500–7500 jobs out of a total of over four million in the metalworking and engineering industries.[47]

The main productivity gains in using FMS come not from direct labour-saving but from more efficient use of capital equipment and from running a third, often unmanned, shift. Indeed, in many cases it is argued by users that retaining human operators in the system has the effect of increasing flexibility and also helping to cope with unexpected problems quickly, hence keeping utilization high.

Occupational structures are changing as a result of the introduction of FMS, with a general cutback in the numbers of unskilled and semi-skilled functions such as loading, unloading, transportation and progress chasing. A consequence of the declining numbers of direct operators, and the need to maintain high levels of utilization and flexibility, is that skill requirements change. Not only is there a need for higher levels of skill but also for greater breadth and flexibility to move between and across skill boundaries.[48] Multiple skilling at all levels is increasing in importance, and the portfolio of relevant skills is also changing, with increasing emphasis on preventative action rather than direct intervention, on diagnosis and problem-solving and on planning and programming.

For example, dependence on maintenance crews can be critical; a point which is illustrated well by Handke in his analysis of the MBB FMS at Augsburg. Here, analysis of over 6000 hours of operation of one of the largest FMS installations in Europe (involving some 30 machine tools and related equipment for the production of components for the Tornado aircraft) suggested that maintenance skills were crucial to minimizing downtime. These skills were not just in repair but also in fault-finding and problem-solving; the overall conclusion of the study was that 'the more complex and automated the systems were, the higher the skills levels of maintenance specialists had to be to achieve reasonable failure rates and implement facility improvements'.[49]

Work organization for flexibility

There is also growing evidence to support the view expressed in the previous chapter that technologies such as FMS do not, of themselves, determine the patterns of skill or work organization associated with the technology. Rather, it is a matter of negotiated choices between key actors in the process – managers, trade unions, workers, and so on. However, there is a strong *a priori* case for arguing that rigid, highly specialized and differentiated working arrangements (such as might be found in a 'traditional' mass production facility) may be inappropriate for conditions where investments are being made to increase flexibility and responsiveness.

Analysis of FMS installations in the Federal Republic of Germany and comparison with data from US and French systems[50] suggested that most users of FMS attempted to implement them within traditional forms of work organization. At the level of substitution – that is, improving local operating efficiency – this arrangement was reasonably effective, but attempts at exploiting the broader potential for flexibility within such systems posed a number of problems.

In general, the desired level of flexibility in operation, in terms of accommodating sudden schedule or design changes (with their implications for resetting, and so on), can only be accommodated effectively if the skills and personnel for these tasks are available within the cell. Dependence on outside functions, with other, competing priorities for this support, may well neutralize any flexibility which investment in physical equipment may offer.

This opens up the question of how far industry is able to shift to more flexible working practices. Up to now, our discussion of how firms can achieve flexibility has focused on the physical technology – machines, computers, robots, and so on. But the pattern of work organization and deployment of staff is a critical determinant of overall flexibility, and one which has been undergoing major changes in recent years. As Atkinson puts it, '. . . we are beginning to witness important changes to orthodox ideas about work organisation and the deployment of labour . . . what is unmistakeable is a managerial imperative to secure more flexibility from the labour force and a readiness to exploit new ways of achieving it'.[51] Part of this pressure comes from the need to utilize increasingly expensive plant more efficiently; for example, by extended three-shift running. With the increasing pressures on skills and overall demographic changes reducing the supply of labour further, there is a greater incentive to develop more flexibility about deployment.

In his study of 'the flexible firm', Atkinson identifies three basic kinds of flexibility in this context:

- *functional*, relating to the ease with which workers can shift between tasks in response to changing technology, markets, and so on

- *numerical*, relating to the ease with which numbers of workers can be adjusted and matched to fluctuations in demand

- *financial*, the extent to which the structure of pay supports the above

In a survey by ACAS, the UK's principal industrial relations arbitration agency, estimates suggest that up to 70 per cent of employers use numerical flexibility (mainly through employing part-time labour in some capacity), 25 per cent have moved towards some form of functional flexibility (such as production workers doing routine maintenance tasks) and a further 25 per cent have moved from 'rate for the job' pay systems to some other, more flexible financial package.[52]

Overall, the need for greater flexibility is calling into question some of the basis principles of traditional production work organization, moving away from rigid payment systems and high division of labour. Research in the UK, for example, has begun to demonstrate the importance of rethinking operator roles within advanced manufacturing systems. In work on small flexible cells (based on an analysis of the causes of system downtime) the researchers found that, in addition to a deskilled machine minding role, there was a need for a highly skilled 'operator midwife' role which involved intervention when problems arose in connection with the largely automated control system. It is important to note that the objective in such systems moves from one in which labour is seen as a necessary evil and as a cost item, to be reduced or eliminated wherever possible, to one in which it is seen as being an important aid to keeping the utilization of the system high – and thus recovering its high capital costs. [53]

In another study, Hirsch-Kreiasen and colleagues in West Germany identified some key work organization design changes which were used as the basis for a parallel organizational and technical change project involving FMS.[54] Traditional work organization in manufacturing involves a high degree of division, both vertically and horizontally, of labour. Vertically, tasks are broken down according to a clear hierarchy: the foreman (scheduling and local changes to plans); the technician (programming and systems control); the skilled worker (setting, maintenance, some programming, and so on); the semi-skilled worker (some operating); and the unskilled worker (loading/unloading). By contrast, FMS operation implies workgroup flexibility, with the team responsible for most if not all aspects. Essentially, the process is one of bringing into the group or cell activities which were hitherto external to it.

The ISF project involved a major manufacturer of gearboxes, producing 16 000 gearboxes each month, mainly for the bus, truck and special vehicle markets, in medium- to high-volume batches. The primary motives in investing in FMS were to reduce lead times and inventory, and to provide some learning about the implications, in order to make more extensive use of AMT in the future. There was also a deliberate attempt to improve the quality of working life, and this last activity was supported by a government grant.

The first machine tools were installed in 1981 and the system became fully operational in 1984. It is designed to handle all 'soft' machining; that is, all necessary operational phases before the final process of hardening the workpieces. Configuration is based on 14 machine tools grouped into 13 machining cells, with one robot and three workpiece carrier stations each. The fourteenth cell is a central load/unload station with a robot. Component feed is by overhead gantry crane and central workpiece storage and retrieval. The whole system is under the control of a mainframe computer. The design did not provide for full automation, but made allowance for a team of six workers per shift on a basis of two shifts a day.

On the basis of some operational experience, a number of changes have been made; for example, to the range of parts actually running on the system. Forecasts for the market demand for some parts proved to be over-optimistic and other parts have been removed from manual machining elsewhere in the plant to run on the FMS. This necessitated some reconfiguration – but also proved the robust nature of the design and flexibility of the system. Software development took longer than expected, and the the levels of automatic control and complexity were reduced because of the high risk of software and system failure, allowing more manual intervention.

The original approach taken to work organization design was basically to follow the team approach, but experience has taught some lessons. Two teams now operate in significantly different modes. Both are heavily dependent on the role of system leader, and these individuals are graded higher than the rest of the group. In Team A three workers are deployed to each side of the FMS, and they have a free choice about who takes responsibility for the various task requirements. After eight weeks they change sides. In contrast, Team B have split work into four areas and a worker is responsible for each; they change around on a four-weekly basis. There is also an overall system foreman, who works on the day shift only and is responsible for procurement and general contacts with other departments.

Overall, the experiment has demonstrated that by bringing into the cell teams able to carry out the basic functions without reference to external groups, and by minimizing hierarchical divisions between workers, considerable flexibility could be built in.

Similar patterns can be found in Sweden. For example, Haywood reports that in one case established customers with urgent component requirements were allowed to by-pass both managers and supervisors and deal directly with the appropriate shop-floor worker to effect relevant changes. In another case operators working with an FMS were able to by-pass the normal structures and contact the responsible shop-floor worker in the supplying foundry directly to arrange for replacement castings. This resulted in replacement components being available within hours, rather than the days which would have been required by moving through the formal system.[55]. Finne describes similar experiences in flexible assembly plants in Norway.[56] Alternative approaches to alternative design strategies have also been embodied in the current ESPRIT project, which aims to develop 'human centred' computer integrated systems.[57]

It is important to note that work organization within flexible manu-facturing systems does not always follow such a 'progressive' path. In another study, Jones and Scott compared work organization in the UK and the US, and found strong support for the view that there is *choice* in the way in which management decide to implement flexible manufacturing technology. That is, there is nothing in the technology itself which dictates a particular form of organization or distribution of skillsand so on – these are matters for choice or negotiation. They found cases in which the systems were used in novel ways – which might be called 'post-Fordist' – effectively breaking away from the traditional models inherited from the earlier manufacturing philosophy. But they also found examples of what might be called 'neo-Taylorism', in which greater horizontal autonomy was given but within a context which still stressed strong vertical control; and of 'neo-Fordism', in which some devolution of autonomy to the manufacturing cell had taken place, but within which emphasis was still placed on deskilling and task fragmentation. In other words, flexible manufacturing technology can be used in to reinforce what went before, or to extend and develop it to other models which may be more appropriate.[58]

Flexibility in industrial relations

Unlike many investments in new technology, the major industrial relations issues surrounding FMS have relatively little to do with redundancy: rather, they are concerned with changes in working practices as workers are pushed towards being more flexible. As Ebel points out, sophisticated control over processes may well be incompatible with the informal processes which workers use to achieve high average performance but to retain considerable control over time and stress at work.[59]

The changes required by alternative working arrangements to promote flexibility have certainly not always been received without opposition. Attempts to secure functional flexibility, for example, imply a major reconfiguration of traditional representation and bargaining patterns, such as the shift to single-union worker representation. This has led, on occasions, to resistance and protracted negotiation with trade unions, especially in countries such as the UK with a tradition of multi-union representation.

Similarly, experiments in alternative working hours have not always been greeted with enthusiasm. Although the general push, in the Federal Republic of Germany has been towards shorter working hours, the case of BMW's new Regensburg car plant is an example of concern being focused on attempts to obtain greater flexiblity in attendance. The plant is designed to give high capital utilization by working 54 hours per week, which is to be achieved through a six-day week of nine-hour shifts. In order to secure this, BMW introduced a novel approach based on allocating 1.5 workers to each workplace, and employing them for a four-day, 35-hour working week. The main resistance to the scheme has come because of the need to work on Saturdays which, the union IG Metall argue, may mean more free time but only at anti-social times.[60]

Industrial relations issues are not confined to the union/management line but can also be found within the trade union/worker representative side itself. The issue of demarcation emerges as a consequence of moves towards greater functional flexibility, and can lead to inter-union conflict. For example, Clegg and Kemp report a problem with a large FMS installation, the implementation of which was delayed by a dispute between two unions over who should be responsible for programming the system.[61]

Flexible manufacturing in flexible organizations

A point made in many studies of flexible manufacturing is that many of the benefits come not from the technology but from the changes which the introduction of the technology forced firms to make in their organization around it. Firms seeking reduced inventories, shorter lead times and higher quality find that their feasibility studies highlight areas of potential organizational as well as technological change. Once implemented, they may find that they no longer need complex integrated systems such as FMS.

A detailed discussion of issues of organizational change raised by AMT can be found in chapter 12. In addition to the changes in skills and work organization mentioned above, some of the key elements include the following.

Group technology

Traditionally, factories have been laid out in a functional arrangement, in which all the tools and equipment associated with a particular activity are grouped together. Products visit these different functional areas during the course of their manufacture. There are strong arguments which are traditionally used to support this approach: for example, it allows for flexibility in the event of breakdown of machinery and it provides a high degree of control over each function. But it has a number of disadvantages, notably the problem of scheduling to cope with unexpected breakdowns and multiple small batches, the pressures for high utilization (functional layouts suit large-batch, standardized manufacture best) and the difficulty of transporting and handling.

An alternative which is receiving considerable attention in the context of flexible manufacturing is to concentrate all the different resources needed to make a particular product or family of similar products into a single manufacturing cell or module. Group technology of this kind is not a new concept – indeed, it was used extensively during the Second World War in the aircraft industry – but it is particularly suited to small-batch, high-variety manufacturing. Group technology requires a number of changes in order to work effectively: rationalization of the product range (together with fixtures and tooling associated with it) so that logical and related families can be identified; and flexible working groups which have functional and numerical flexibility and flexible equipment.

A variant on this theme is the concept of 'factories within the factory', in which flexible cells are equipped with all the functional support and facilities to operate in a highly autonomous way. For example, in the Ulm plant of AEG, a 'mini-factory' has been installed for complex machining of aerospace parts. Here all the activities associated with their manufacture, including maintenance, quality management, purchasing, production control, and so on, have been devolved to the operating unit, which is based around a small number of highly integrated manufacturing cells.[62]

Functional integration

As technology brings different areas of the firm together, it becomes important to ensure that the problems of interdepartmental boundaries are minimized. In some cases this may lead to the creation of new roles or groups, either on a temporary (task force, project team) or on a permanent basis. For example, there is the need – itself facilitated by moves towards CAD/CAM linkages within firms – for the design and production departments to work closely together to develop products which are suitable for manufacture on integrated flexible systems.

Such a 'design for manufacture' philosophy is of particular significance in the flexible assembly automation field, for example, where small modifications to the design of an item can eliminate the need for complex manipulation or operations within an automated system. In one notable FMS case which we examined, redesign of the product led to a reduction in the number of operations (handling and machining) from 47 to 15, with consequent improvements in both productivity and quality.[63] Tidd also draws attention to this factor, suggesting that greater emphasis on design for manufacture is a key feature in the greater flexibility found in Japanese robot applications compared with their UK counterparts.[64]

The essence of such functional integration is not to eliminate specialist skills, but to bring them to bear in a coordinated fashion on the problems of designing, producing and selling products. The intention is to create a single-system view of the process, rather than one with many parochial boundaries across which there is little interchange. The opportunities opened up by IT for computer-integrated manufacturing, exploiting what have been termed 'economies of communication', depend critically on the effectiveness with which the different functions, now linked by computer, can relate to one another. These benefits can quickly become costs:

The systemic aspect of CIM generates economies of communication connected with time savings for all phases of operations, in particular handling of WIP, stock programming, designing, writing specifications for and testing new products as well as Quality Control during production . . . but it is crucial to minimise adjustment times and to avoid manipulation of the information received . . . it is a matter of taking into account the comparative costs of planning, adapting and monitoring several tasks in different organisational structures of the firm.[65]

Vertical integration

In the same way as the integration of technologies requires closer functional integration, so it implies flatter hierarchies and greater vertical integration in the organizational structure. In order to exploit the full benefits of a rapidly responsive and flexible system it may be necessary to create a managerial decision-making structure which is closely involved with the shop floor, and which has a high degree of delegated autonomy. One approach to this is the setting up of semi-autonomous business units, concentrating not only the necessary production facilities and support associated with a particular product family, but also the relevant business and financial functions.

Beyond this, there is the more general need for changes in organization culture, the set of beliefs and norms about 'the way things are and the way we do things around here'. The challenges posed by integrating technologies will require new ways of thinking about how to organize to make best use of them. But the organizational ability to exploit the technologies successfully will depend on how far the prevailing culture is open to change. Traditionally production has been characterized by a culture which emphasizes such things as stability, bureaucratic procedure (as in 'doing things by the book'), specialization and division of responsibility. Although such a culture (which Burns and Stalker labelled 'mechanistic') was traditionally well suited to the demands of production in a stable environment, it is less well suited in one characterized by fluctuating demands in the marketplace, where agility, responsiveness and flexibility are the key factors associated with success. Consequently, there is a need to develop ways of moving towards a more open and flexible culture in production, one which Burns and Stalker term 'organic'. And this may again have implications for structures, methods and processes within the firm.[66]

5.8 MANAGING FLEXIBLE MANUFACTURING TECHNOLOGY

Changes such as moving to group technology/product family approaches, increasing functional integration, developing the skills base of the organization and so on are all important parts of creating a more flexible and responsive organization. But we need to be aware of the questions this raises about how to make such changes effectively; and this requires skill in identifying and implementing what can often be radical alternatives to conventional approaches. The supply side now offers a range of technological solutions, so that it is more or less possible for any firm to obtain the physical technology for flexible manufacturing. It is therefore the way in which that firm *manages* technology which is going to determine the extent of the competitive advantage that can be obtained from this investment.

Two themes central to this process will be: how to exploit the strategic (as opposed to the simply operational) advantages of flexibility; and how to exploit the wide range of choice (about organizational as well as technological characteristics) which is becoming available.

As we saw earlier, flexibility can be seen as a response to uncertainty in the environment. In table 5.6, Gerwin suggests ways in which the different kinds of uncertainty can be dealt with, not only by the technology of flexible manufacturing, but also by the effective use of existing organizational resources.[67] The key message here is not one of either/or but rather of what can be achieved with an effective combination of the two.

Flexibility type	Workforce characteristics	Equipment
Mix	Varied skills	Low degree of specialization
Changeover	Ability to learn new operating skills	Little hard automation and use of programmable technology
Modification	Ability to modify operating procedures quickly	Standard fixturing
Rerouting	Group technology and working arrangements	Redundancy in equipment
Volume	Varied skills available external to the line (can be used elsewhere)	High adjustability of capacity
Material	Varied skills in maintenance and defect detection and correction – available through group working	Adjustment and correction mechanisms
Sequencing	Varied skills available within the line team – team or foreman controls balancing	Fast set-up equipment

Table 5.6   Sources of flexibility in organizations

There is a growing perception of the need for alternative models of production organization and management to support effective use of flexible manufacturing technology. At one level this can be seen as mirroring the basic elements of the architecture of the computer systems being installed, with emphasis on networking rather than hierarchical control, on delega-tion and local autonomy rather than centralization, on communication rather than supervisory control, and on integration. But it is important not to forget the extent to which some of the organisational factors are interrelated. For example, education and training are needed, not only to provide the necessary skills base to support the technology but also to permit alternative forms of work organization. Attempts at devolving autonomy and responsibility to the shop floor will only succeed if there is the right skill base present to take up the challenge. Consequently, any attempt to develop the organization in parallel with the technical systems will require a similar commitment and investment.

It is significant to note that several references have already been made to the Japanese experience, contrasting it with that in various Western nations. It appears that Japanese firms are generally able to use advanced manufacturing technologies of various kinds more flexibly and more productively than their competitors – even when we measure flexibility along its many dimensions. One strong argument – which we will take up later – is that this is largely due to the fact that the need for the alternative approaches to production organization and management which we have identified has already been taken on board. Many of the characteristics of 'Japanese manufacturing techniques' are precisely those which provide support for the effective use of flexible manufacturing technology.

5.9 SUMMARY

To conclude, it can be argued that we are moving into a new manufacturing era as we approach the twenty-first century, in which the qualities of flexibility and agility will be at a premium in manufacturing and services. There is also a growing recognition that the modes of manufacturing which were appropriate to an earlier era of mass production are likely to become less appropriate as we move forward, and so there is a need to look for alternative models to succeed those of Ford and Taylor.

At the same time the availability of powerful information-based technologies is also developing hand in hand with a recognition of the increasingly important role of knowledge (both in terms of direct skills and also in terms of tacit knowledge) in the manufacturing context. Arguably, we are moving from an emphasis on capital intensity to a period in which capital and knowledge accumulation takes place, and in which we will be less concerned with labour as a cost and more so with its effective exploitation as a key competitive resource.

'Technology' as defined in the dictionary, is not an arrangement of machines and computer hardware and software, but a much more holistic system representing 'the useful arts of manufacture'. This implies that we should look for optimal arrangements which bring together the sophisticated tools of computers, machine tools and their like with our enormously flexible and adaptable human resources. Within this framework the successful flexible manufacturing system of the future is therefore likely to be characterized by a high degree of interaction between human and computer, exploiting the best features of both. In essence, obtaining a flexible advantage in manufacturing has relatively little to do with *buying* a flexible manufacturing system, and a great deal to do with *becoming* a flexible manufacturing organization.

Notes

1   *Metalworking News*, 1 August 1988, pp.38–49.
2   D. Gerwin, 'An agenda for research on the flexibility of manufacturing processes', *International Journal of Operations and Production Management*, 7, (1) (1986).
3   S. H. Lim, *An Organisational Survey of Flexibility in Britain* (Scottish Business School, Glasgow, 1986).
4   J. Krafcik, 'The triumph of lean manufacturing,' *Sloan Management Review*, 30 (1) (1988), pp.41–51.
5   J. Tidd, *Flexible Manufacturing Technology and International Competitiveness* (Frances Pinter, London, 1991, forthcoming).
6   R. Camagni, 'The flexible automation trajectory: the Italian case', paper presented to EEC Symposium on *New Production Systems: Implications for Work and Training in the Factory of the Future*, Torino, 2–4 July 1986.
7   For example, see H. Boer, M. Hill and K. Krabbendam, (1988) who found in their study of seven FMS installations in the UK, Belgium and the Netherlands dual sets of motives, one group concerned with operational objectives and the other with strategic objectives. (H. Boer, M. Hill and K. Krabbendam, *It is One Thing to Promise and Another to Perform: The Case of FMS Implementation Management*, Working Paper, University of Twente, School of Management Studies, December, 1988.
8   For a detailed review of these concepts, see N. Slack, 'Flexibility of manufacturing systems', *International Journal of Operations and Production Management*, 7 (4) (1987) pp.35–45; and 'Focus on flexibility', in *International Handbook of Production and Operations Management*, ed. R. Wild (Cassell, London, 1989).
9   Tidd, *Flexible Manufacturing*.
10 W. Abernathy, *The Productivity Dilemma: Roadblock to Innovation in the Automobile Industry* (Johns Hopkins University Press, Baltimore, 1977).
11 A de Meyer, *Flexibility – the Next Competitive Battle*, INSEAD Working Paper, WP/86/31, Fontainebleau, 1986.
12 Cited in Tidd, *Flexible Manufacturing*.
13 M. Piore and C. Sabel, *The Second Industrial Divide* (Basic Books, New York, 1982).
14 Slack, 'Flexibility of manufacturing systems'.
15 R. Jaikumar, 'Post-industrial manufacturing', *Harvard Business Review*, November 1986, pp.69–76.
16 S. Jacobsson and C. Edquist, *Flexible Automation*, (Basil Blackwell, Oxford, 1986).
17 An excellent review of the development of NC technology and beyond can be found in Jacobsson and Edquist, *Flexible Automation*.

18 J. Baranson, *Robots in Manufacturing* (Lomond, Maryland, 1983).

19 'Planning for automation success in the food industry', *Food Europe*, supplement to *Food Processing*, May/June 1987.

20 A fuller description of this process appears in A. Bhalla et al, *Blending of New and Traditional Technologies* (Tycooly International, Dublin, 1984), see ch.5.

21 After extensive debate, the main patent for Molins System 24 and the title 'flexible manufacturing system' was finally granted in 1983. This covered the installation of computer-controlled machine tools linked together by some form of automated component transfer and backed by component and tool stores. In its final form it covers 254 separate claims for FMS layouts based on 90 primary combinations of machine tools, computer, transport and stores systems. The company argues that this patent holds for the present generation of FMS equipment and has approached 200 US firms to demand a $100 000 licence fee and a 2 per cent royalty on sales! (*The Engineer*, 19 June 1986, p.10.)

22 J. Ranta and I. Tchijov, *Economics and Success Factors of FMSs*, IIASA Discussion paper, Vienna, 1988.

23 UN/ECE, *Recent Trends in Flexible Manufacturing*, (United Nations Economic Commission for Europe, Geneva, 1986).

24 J. Ranta, *Trends and Impacts of CIM*, IIASA WP-89-1, International Institute for Applied Systems Analysis, Vienna, 1989.

25 Ranta and Tchijov, *Economics and Success Factors of FMS*.

26 Cross International, quoted in *Automation*, October 1987.

27 Ranta and Tchijov, *Economics and Success Factors of FMS*.

28 International Federation of Robotics, cited in *Financial Times*, 5 December 1989.

29 J. Northcott, *Robots in British Industry*, (Policy Studies Institute, London, 1988).

30 U.S. Department of Commerce, *Manufacturing Technology*, Report, Bureau of the Census, Washington D.C., 1989.

31 J. Bessant and W. Haywood, 'Flexible manufacturing in Europe', *European Management Journal*, 6 (2), pp.134–55 (1988), more detailed reports on these studies can be found in the following Occasional Papers: (1) The introduction of FMS as an example of CIM (1985); (2) FMS and the small/medium sized firm (1986); (3) FMS in Sweden (1987) – all published by the Centre for Business Research of Brighton Polytechnic.

32 M. Ollus and J. Mieskinen, 'Bases for flexibility in a small country', in *Trends and Impacts of CIM*, ed. J. Ranta (IIASA, Vienna, 1989), pp.355–78.

33 FIR, *Einsatz von Flexiblen Fertigungszellen – Bestandsaufnahme*, Forschungs-institut für Rationalisierung an der Rheinisch-Westfälischen TH, Aachen, 1987.

34 It should be noted that in a number of cases direct comparisons against previous ways of working were inappropriate because plant represented new investment for new products.

35 M. Tombak and A. de Meyer, 'Flexibility and FMS: an empirical analysis', *IEEE Transcations on Engineering Management*, 35 (2) (1988), pp.101–7, .

36 Boer et al, *It is One Thing to Promise*.

37 M. Graham, A tale of two FMSs, in *Managing Advanced Manufacturing Technology*, ed. C. Voss (IFS Publications, Kempston, 1986).

38 G. Margirier, M. Holland and A. Rosanvallon, *L'autonomisation avancée de la production dans les activités d'usinage*, FOP No 124, FAST Programme of the EEC, Brussels, November 1986.

39 S. Mori, 'Trends and problems of CIM in Japanese manufacturing industries', in Ranta, *Trends and Impacts of CIM*, pp.379–402.

40 H. Rush, J. Bessant and K. Hoffman, *Evaluation of the FMS Scheme*, Report to the Department of Industry, Centre for Business Research, Brighton Business School, January 1990 (mimeo).

41 In interviews the emphasis was repeatedly placed on FMS as a 'strategic' technology and not just one more increment of technological capacity.

42 Similar experiences are reported in the case of CAD systems, for example. See P. Senker and E. Arnold, *Designing the Future*, Occasional Paper 9, Engineering Industry Training Board, Watford, 1984, for a discussion of this.

43 Margirier et al., *L'autonomisation avancée*.

44 Jaikumar, 'Post-industrial manufacturing'.

45 For example, see Tidd, *Flexible Manufacturing*.

46 P. Dempsey, 'New corporate perspectives on FMS', in *Proceedings of Second International Conference on Flexible Manufacturing Systems*, ed. K. Rathmill (IFS Publications, Kempston, 1982).

47 K. Ebel, 'Social and labour implications of FMS', *International Labour Review*, 124(2) (1985), pp.365–70.

48 P. Senker, *Towards the Automatic Factory* (IFS Publications, Kempston, 1985).

49 G Handke, 'Design and use of flexible automated manufacturing systems', in *Proceedings of First International Conference on Flexible Manufacturing Systems*, ed. K. Rathmill (IFS Publications, Kempston, 1981).

50 Based on Margirier, et al., *L'autonomisation avancée;* and D. Gerwin and M. Blumberg, 'Coping with advanced manufacturing technology', *Journal of Occupational Behaviour*, 5 (1984), pp.113–30,

51 J. Atkinson, *The Flexible Firm*, (Institute of Manpower Studies, Sussex University, 1984).

52 ACAS, *Labour Flexibility in Britain – the 1987 ACAS Survey* Occasional Paper 41, ACAS, London, 1988.

53 T. Wall and N. Kemp,'The nature and implications of advanced manufacturing technology', in *The Human Side of Advanced Manufacturing Technology*, ed. T. Wall et al. (John Wiley, Chichester, 1987).

54 H. Hirsch-Kreinsen, 'Skilled production work in an FMS – alternative work organisation in a mechanical engineering plant', in *Human Factors in Systems Design*, ed. F. Prakke, report to DG-V, EEC, TNO, Apeldoorn, 1987.

55 W. Haywood and J. Bessant, *The Swedish Use of FMS*, Occasional Paper, Centre for Business Research, Brighton Polytechnic, 1986.

56 H. Finne , 'Human factors in the design and implementation of a system for flexible automated assembly of electronic convector heaters', in Prakke, *Human Factors.*

57 J. Corbett, 'Design for human–machine interfaces', in *New Technology and Manufacturing Management*, ed. M. Warner et al, (John Wiley, Chichester, 1990).

58 B. Jones and P. Scott, 'Working the system: a comparison of management of work roles in American and British flexible manufacturing systems', in *Managing Advanced Manufacturing Technology*, ed. C. Voss (IFS Publications, Kempston, 1986).

59 K. Ebel, 'Social and labour implications of FMS'.

60 'IG Metall threatened by BMW's quest for more workers to do less work more often', *Financial Times*, 8 August 1989.

61 C. Clegg and N. Kemp, 'Information technology personnel – where are you?', *Personnel Review*, 15 (1) (1986), pp.8–15.

62 M. Chakrobarty, 'Implementation of mechanised manufacturing cells in existing production as an economic alternative to FMS', in *Proceedings of Third International Conference on Flexible Manufacturing Systems* (IFS Publications, Kempston, 1985).

63 J. Bessant and W. Haywood, *The Introduction of Flexible Manufacturing Systems as an Example of Computer-integrated Manufacturing*, Occasional Paper, Centre for Business Research, Brighton Polytechnic, 1985.

64 Tidd, *Flexible Manufacturing*.

65 L. Filippini and A. Rovetta, 'Economic aspects in factory automation in relation to system flexibility', paper presented at a conference on Methods in OR and FMS, CISM, Udine, Italy, 5–9 October 1987.

66 T. Burns and G. Stalker, *The Management of Innovation*, (Tavistock, London, 1961).

67 D. Gerwin, 'An agenda for research on the flexibility of manufacturing processes', *International Journal of Production and Operations Management*, 7 (1) (1986), pp.38–49.

6 Keeping Track of it All: Computer-aided Production Management

6.1 INTRODUCTION

At the core of any manufacturing company lies an information system. Although the business may appear to involve physical flows of components, materials, products, and so on, this is underpinned by a complex set of information flows. Managing these effectively involves making sure the right information is available at the right time to the right people and failure to achieve this will lead to a variety of problems such as delayed deliveries, overstocking and poor inventory control.

Not surprisingly, this area represents a prime target for AMT, especially in the application of computer-based information systems. Just a few of the pieces of information which need to be collected, communicated and coordinated in a typical factory are listed below:

- sales orders

- purchasing of raw materials and components

- orders for in-house manufacture of necessary components and sub-assemblies

- instructions for tool and fixture preparation and management

- management of transportation within the works

- machine loading and scheduling against capacity

- production planning for the factory as a whole

- production control over what is actually happening

- cost accounting over each stage

- documentation and certification

- invoicing and financial tracking

- quality and test procedures and results

6.2   NEED PULL

As far as forces pulling new technology through in the area of production management information are concerned, the problem is not a new one. The ideal in production management has always been to have the right amount of information available at the right place at the right time. So, for example, the purchasing manager needs to know exactly what to order at the time he is placing the order to suppliers so as to be within their lead times, but without filling his stores with costly inventory for too long. The production manager needs to know exactly how much product he has to make so as to be able to optimize the utilisation of his machines and labour. The customer service manager needs to know the exact status of a customer order so as to be able to answer queries as and when they come in and to be able to quote delivery dates which are achievable.

The problem lies not in the availability of such information but in its being shared across the organization in a usable form. Within each functional area the availability is generally good: the stores manager knows how much is in stock and what has been ordered, the production manager knows the status of his machinery and labour, and so on. This information may not always be held on record – it may be in the head of one manager, on the back of an envelope for another, or in a complex card index for a third. But it is there and, generally speaking, the quality and integrity of such local information is good.

Communicating the right information to the right person at the right time is more complex, however, especially when some form of processing and integration of information from more than one source is also needed. For example, the materials manager would like to know from the marketing department how many products he will need, and from the production manager how many he can make, so as to arrive at a realistic purchase order which minimizes costs of inventory carrying while ensuring that production can take place.

The problem is further compounded by the uncertainties involved. The marketing manager can only make a 'guesstimate' ahead of time about how many he will need and, as with all forecasting, even the best guess is still prone to error. Similarly, the production manager cannot accurately predict the status of his shop ahead of time: he cannot anticipate that his key machine, through which everything has to pass, will break down the night before a big production run, thereby delaying all batches for the following week and necessitating a switch in production to the manufacture of something else, for which that machine is not needed.

Therefore, the need is for a system which goes beyond simply supporting local-level information processing. It needs to be one which can not only handle basic 'number crunching', but which can also integrate information for different users, drawing on a common database. It also needs to be able to react quickly enough to changes in the environment within or outside the works, and recalculate fast enough to present alternative information if necessary. Ideally, it should be able to store data over time and learn from this data, so that its future behaviour will anticipate some of the problems of poor forecasting, and make allowances – a self-optimizing system. It should be able to generate reports to suit a wide variety of different users. And it should be available for use by a wide variety of different users, in different locations and with different backgrounds and experience.

Broadly speaking, this is the specification for a computer-aided production management (CAPM) system, which has evolved over time. The key characteristic of all of this is that it can only realistically be handled by IT. The volumes of data, the timescale, the communication, and so on, would all make a manual system impossible to implement. As we saw earlier in the case of FMS the pent-up needs for such a system have contributed extensively to the birth of CAPM systems. And once they emerged in the 1960s and 1970s, the particular needs of different departments, for learning and feedback and – more recently – for wider access to 'user friendly' systems, have all shaped the evolution of this technology. Simultaneously, the presence of different types of firm with very different coordination needs has also been a powerful force, so that there are CAPM systems for the process industries, for the sub-contracting industries, the volume producers, and so on.

6.3   TECHNOLOGY PUSH

In a manual information system, such as we might have found in the 1950s, all of the information listed in section 6.1 would be collected separately and then used in integrated form where appropriate. For example, the stores would be managed by a record system which tracked what was in stock and, using some form of re-order system, would issue orders to ensure that there was always sufficient available for the manufacturing programme. The sales orders would be processed by an army of clerks; and the information about what was required would be passed on to the production planners, who would combine the information that they had collected about the available machines and their capacity with the information from the design department about what each product involved in terms of raw materials: from this they would be able to place orders for purchasing these from outside, or for manufacture within their own facilities.

The point about all of these activities is that the individual information-processing tasks are simple and repetitive – essentially basic arithmetic involving counting, multiplication and division. For example, the basis of machine scheduling is to ensure that each machine is loaded as efficiently as possible. In order to do this, the jobs to be done per week or per day are divided by the number of hours it takes to complete each one. Or, again, the materials requirement for a week's production involves working out how many of each component are needed per product and multiplying by the number expected to be made. The calculation of how much to order is then basically a subtraction operation: what is needed less what is already in the stores.

Of course, although the basic information tasks are very simple, the sheer numbers and interrelationships, coupled with the inevitable uncertainties of manufacturing life (machine breakdowns, late deliveries, and so on) mean that production management is complex in practice. But at heart it is a process well suited to the application of computers in their basic 'number crunching' mode.

Not surprisingly, the idea of CAPM is attractive. By harnessing the power of computers it should be possible to manage even highly complex production to optimum efficiency – with minimum stock, reliable delivery, and so on.

CAPM was certainly seen as one of the major potential areas of application for computers in the 1960s, but at that time machines were expensive and cumbersome and confined to the larger firms only. More importantly, access to them was via a complex set of languages and communication devices such as punched cards, which were usually the property of the 'high priesthood' of the Data Processing Department. Although the software made it theoretically possible to integrate information from many different sources and to provide a shared system, in practice it was often very difficult to access that information in a useful form, or when it was urgently needed. Programs were often relatively slow, so runs were carried out on a weekly basis or at even longer intervals, using quiet times (often weekends) in order not to monopolize the company's main computer.

By the 1970s the emergence of minicomputers had led to a much wider spread of application of computers in manufacturing, with many more firms able to take such technology on board. Languages had become a little more penetrable, engineers were also being trained to use the tools and the DP stranglehold was beginning to show signs of strain. Most importantly, lower-cost minicomputers also meant that it was economic to buy more terminals and to distribute access to the computer throughout the organization at least a little more widely. This meant that users in different areas could start to have access to computer power for their own information activities and that – potentially, at least – the option existed for integration, for sharing of that information by others who also needed it for their calculations.

The most significant development at this stage was the emergence and refinement of integrated CAPM packages. A typical example of this is the software for Materials Requirements Planning (MRP). These systems, often available in modular form, enabled integration of information flows between separate functional areas enagaged in making sure that there were sufficient raw materials available to manufacture the week's production.

MRP is a way of controlling the process of ensuring the availability of finished products to meet orders, and of the necessary components of manufacture. It is a simple concept, but the practical difficulties of making sure that variations (in customer demand over time, of costs, of uncertainties in manufacturing capacity for components made in-house or of shortages or delays in delivery of bought-in items, and so on) can be catered for without carrying excessively high inventories mean that without the use of computers only very superficial and infrequent MRP can be carried out.

In a computer-based MRP system the computer forges a link between information about production, purchasing and marketing, and provides a management tool with which to identify priorities in purchasing and production, to optimize the use of capacity and to integrate supply and demand. In the longer term it offers ways of improving sales forecasts, reducing inventory levels and improving the purchasing function.

A typical MRP system is illustrated in figure 6.1, and its basic operation is illustrated in figure 6.2 .

The planning process begins with a forecast of what the company expects to sell during a given period. This is converted into a production plan which sets out what will be made, and what the implications will be for the various production resources. The outcome of this process of breaking down the plan into resource requirements is known as the Master Production Schedule (MPS).

The MPS is then used to calculate which components will be required, and this information is compared with what is held in stock to identify what needs to be bought. It does this by referring to a file containing details of all the components needed to make the product in question – the Bill of Materials (BOM) file. Again, the concept behind a BOM file is simple; but the complexity can be illustrated if we consider the sheer number of components and assemblies which might be required for a particular product – a television set, for example. As we begin to break the set down into its component parts, we can see the complexity growing, and by the time we reach the (literal) nuts and bolts level there may be several thousand components involved. So far, this structure only refers to one model: in practice, like most other products, TV sets will be offered in a range of options, each with its complicating influence on the BOM data.

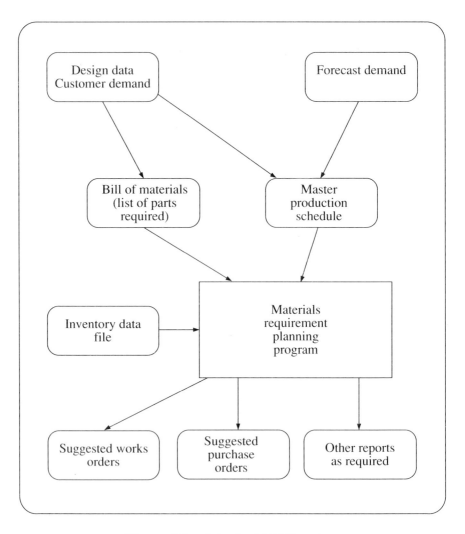

Figure 6.1   A typical MRP system

Maintaining an accurate and usable BOM file is clearly a task better suited to computer-based information processing than to manual work. Computer-based systems offer a number of additional advantages: for example, they can quickly identify commonality between components required for different models, and they can be updated quickly as designs are changed.

The final stage of the system is to use the MPS to generate a detailed Material Requirements Plan which sets out in detail what needs to be bought in, made in-house, and so on. With a manual system, every time the MPS is recalculated major changes would be involved: with a computer-based system it is possible to reduce fluctuations of this kind.

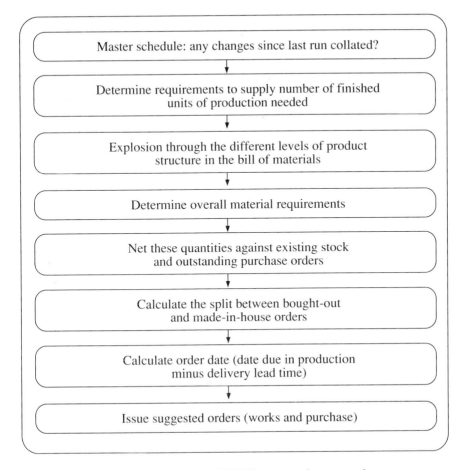

Figure 6.2   A typical MRP system in operation

Although such systems appear to be logical and relatively simple in concept, there have been significant problems in their use since the 1960s, when basic MRP systems were first introduced. Early problems had more to do with organizational and human factors than technological factors, including:

- poor quality data input (because of lack of commitment or even deliberate action). – poor data in the system renders the generated information ineffective or incorrect;

- poor implementation – many systems remained the province of data processing experts and were often imposed upon the rest of the organization;

- lack of commitment from senior management;

- slow in operation (runs could take several hours) and unresponsive to changes

- lack of feedback provision to take account of changes in capacity, order levels, lead times, and so on

- often seen as the responsibility of one department, usually data processing (DP) or stock control, rather than an organization-wide responsibility

- weak links to other aspects of the production process, such as quality control

The emergence of MRP2

MRP grew out of the expansionist US manufacturing climate of the 1960s. The massive domestic market and booming economy led many manufacturers to emphasize high-volume batch manufacture, and under these conditions the main problem was ensuring that sufficient material was available to maintain production. The subsequent changes in economic climate put pressure on firms to make more effective use of capacity and to increase both flexibility and their ability to produce smaller batches economically.

But, by their nature, MRP systems worked best for those firms with little basic variety in product range and with relatively stable patterns of orders and supply. Such systems had diffused widely but it was clear that major changes were needed in order to exploit the full potential of CAPM. By extending the role of MRP beyond simply automating the stock control area, to cover a process which optimized the use of all production resources and the link between production and the rest of the business, the system could become a much more powerful aid to management.

This was the philosophy behind the emergence of Manufacturing Resources Planning (MRP2), which essentially involves closing the loop between supply and demand. In essence, it means that material requirements are not generated without some reference to the implications of orders for capacity needs and the influence of resource constraints. Although originally a proprietary piece of software, MRP2 has become a generic name for a range of integrated CAPM packages which still represent the main option in this area for manufacturing organizations. Versions have been developed to suit particular needs, ranging from small jobbing firms to high-volume process industries, but the core concepts remain the same. The essential features are illustrated in figure 6.3. As a concept, MRP2 was well established by the 1980s. The most significant changes at this stage came in the availability of low-cost terminal power in the form of the PC. Such systems could either be

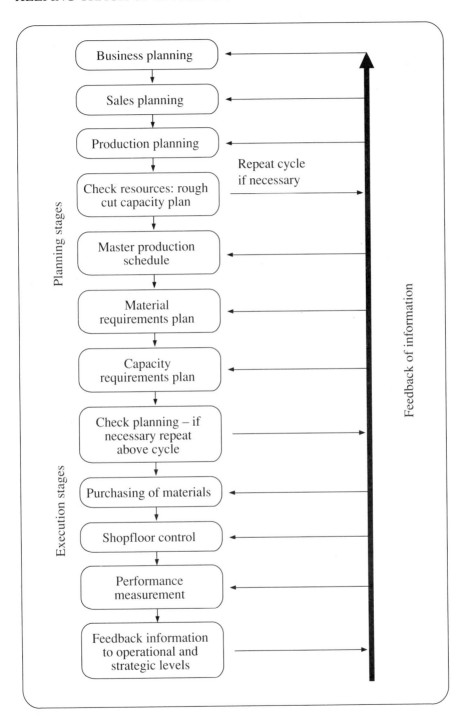

Figure 6.3   MRP2

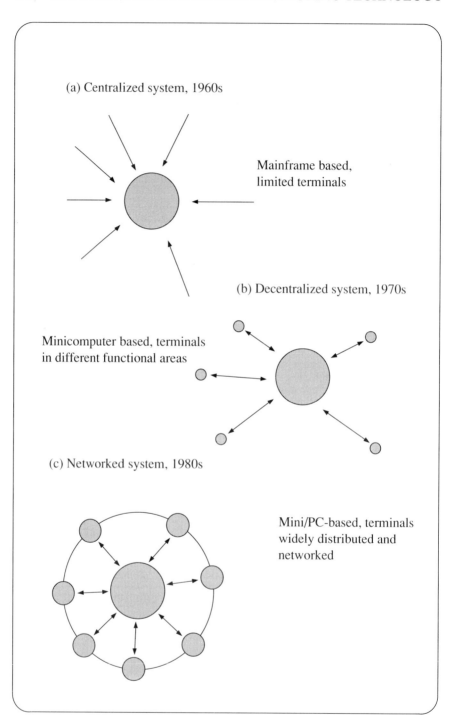

(a) Centralized system, 1960s

Mainframe based,
limited terminals

(b) Decentralized system, 1970s

Minicomputer based, terminals
in different functional areas

(c) Networked system, 1980s

Mini/PC-based, terminals
widely distributed and
networked

Figure 6.4   Cyclic process of development of CAPM

used as intelligent terminals connected to a larger network with a mini-computer at its heart, or they could be used as local stand-alone processors, carrying out discrete information activities such as stores management. The net effect was to diffuse the availability of access – both to put information in and to take it out – to such systems and open up the possibility for even higher levels of integration within the business. A second, emerging feature was faster processing, so that the weekly MRP number crunch became a much faster – and therefore more usable – option. This was made possible by developments in hardware (faster chips and architectures) and software, especially in more structured routines and languages.

The current picture

As we enter the 1990s, the present situation takes us one stage further: we now have not only the networking of PCs linked to minicomputers or mainframes but also the emergence of hand-held devices, from simple bar code readers to sophisticated portable computers. The net effect is that there is now the possibility of collecting even more information and integrating it into the network in real time.

In many ways the pattern of access to production management information has returned to where it started – at the local level. The difference is that such local-level systems, which provide highly specific information for a particular area or activity, are now also part of an integrated network in which inputs from a variety of sources can be shared; and the power of the central processing facility can be used to enhance the range and type of information available. This cyclic process of development over the three generations of CAPM systems is illustrated in figure 6.4.

There are now several hundred different CAPM packages available to suit a wide range of applications; by firm size, industrial sector, and so on. Most offer some form of modular software, in which basic functions such as MRP or the BOM file can be added to so as to provide a customized solution to a particular set of needs, and to offer potential for expansion. A typical set of modules for CAPM is illustrated in figure 6.5.

6.4   DIFFUSION OF THE TECHNOLOGY

With such a wide range of choice on the supply side, and given that effective management of production information has been such a long-standing problem, it is not surprising that the diffusion of CAPM technology is already widespread. In recent years there has been substantial growth in applications within smaller firms, predominantly because of the availability of simpler, low-cost entry-level systems based around PCs.

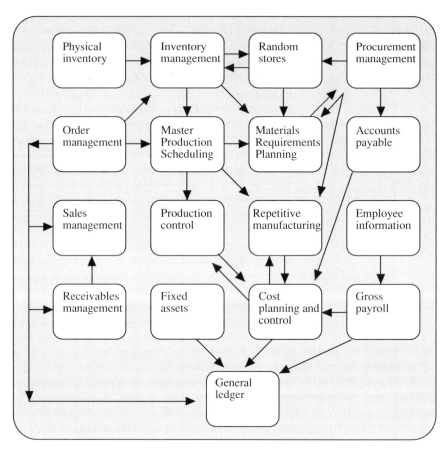

Figure 6.5   A typical CAPM system

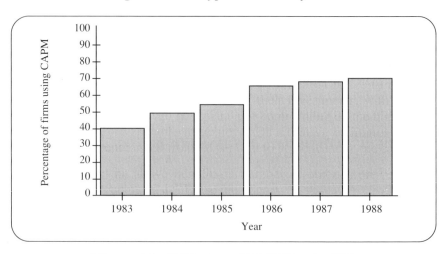

Figure 6.6   Diffusion of CAPM in the UK

One indication of the rapid growth in application in the UK is given in figure 6.6. It draws on statistics published by the *Engineering Computers* magazine, which surveys a statistically representative sample of around 650 firms every year. From this some broad indications of the UK pattern as a whole can be derived and – as the graph shows – there has been substantial growth in recent years in CAPM application.

6.5   BENEFITS

Investment in CAPM should lead to improvements along several dimensions; for example, in lower inventories and shorter lead times, since delays due to shortages of key components would be reduced. Corke[1] suggests the following list of benefits which CAPM offers to manufacturing managers:

- to enable delivery periods to be offered which are short enough not to lose the company valuable business

- to deliver customers' orders by quoted dates

- to utilize the company's resources of plant and manpower to achieve the requested output on time and at the lowest possible cost

- to plan and control the levels of work-in-progress and stock, keeping these to a minimum

- to change the manufacturing capacity as necessary and in time to achieve objectives in the face of a changing volume and mix of business

- to provide systematic planning and control of the procurement of material and its progress through the stages of manufacture, as a background against which to deal with unforeseen events.

There is considerable evidence to support the ability of CAPM systems to achieve these objectives, as highlighted in table 6.1, which details some case studies drawn from the technical press.

6.6   BEYOND MRP AND MRP2

MRP and related systems are now widely available for a variety of firms in all sector, size and price brackets. From the technology push side, the picture is very much one of proliferation of choice, and this has been extended in recent years by the emergence of alternative approaches which combine or compete with CAPM as solutions to the production management problem. These include just-in-time (JIT), optimized production technology (OPT) and a variety of hybrid systems, which it will be useful to examine briefly.

Company	Investment	Inventory saving	Lead time	Other
(a) Cox Pharmaceuticals	£240 000	£500 000 in raw material	—	Inventory static at £3.5 million despite turnover increase from £14 million to £17 million
(b) Astec (audio products)	£200 000	30% cut in work-in-progress	'Weeks to days'	—
(c) Rexel Stationery	£400 000	50% cut, £2 million to £1 million, 75% of this in 13 months	Cut by 30%	Output rose in this period
(d) Heenan Drives	£250 000	Cut by £500 000	—	—
(e) Leyland–DAF	—	Cut from £26 million to £12 million, inventory turns up 5 to 14	Cut from 13 to 6 weeks	Cut costs by 50% – Output up from 120 to 340 trucks per week during this period
(f) Fulcrum	—	Inventory turns up 70%	Cut by 40%	Indirect staff cut by 45%
(g) Ciba–Geigy (Lyons)	£120 000	£65 000 annual savings	—	Staff savings £40 000 per annum

Sources:	(a,b,c) *Works Management*, October, 1988
	(d) *Works Management*, October, 1987
	(e) *Engineering Computers*, January 1989
	(f) *Engineering Computers*, December, 1988
	(g) Trade literature

Table 6.1   Benefits from CAPM investment

Just-in-time systems (which are discussed in detail in chapter 8) are based on an alternative approach to inventory management. They attempt to minimize inventory by only making components available just in time for them to be used – and then apply this principle right along the supply chain. Whereas MRP and similar systems make a forecast of what will be required and then use this information to trigger action by 'pushing' inventory into the system so that it is ready for use, JIT systems do the reverse. They wait until there is a demand – a 'pull' – and only then do they release inventory.

JIT systems have proved to be highly successful in reducing inventory in certain cases because they do not permit it to build up 'just in case' of problems which disrupt the normal production pattern. At the same time there is a higher risk of running out of material unless the whole production and supply process becomes increasingly geared to smooth flow and rapid response. For this reason, as we shall see later, implementation of JIT requires much more than simply changing the approach to inventory management: it needs training, reorganization and a commitment to continuous improvement and problem-solving. JIT also emphasizes the importance of factors such as reduced set-up time, so as to facilitate the manufacture of smaller batches.

JIT works best when batch sizes are fairly high and where there is a fairly predictable market environment: it can run into problems when it has to try to react quickly in uncertain environments unless the manufacturing system is geared up for very rapid response. A number of users have begun to utilize the master schedule development parts of MRP2 packages (that is, their planning functions) to provide the manufacturing plan which can then be executed using a just-in-time approach.

Various 'hybrid' JIT/MRP systems are in use in different firms; for example, the Synchro-MRP system or the Lucas model. In these hybrids the advantages of both approaches are combined to give workable production and materials management; for example, by using JIT 'kanban' cards to pull work through the factory, while using MRP to procure material.

Optimized production technology (OPT) also makes some different assumptions about production to MRP. In this case the basis is an analysis of the production process in terms of bottleneck operations. Everything in the factory is tied to the output speed of the slowest bottleneck, and scheduled back from that point using a complex mathematical algorithm. The aim is then to keep the bottleneck fed continuously, but to ensure that operations upstream of it only produce just in time for the bottleneck operation to use their output – any faster rate of production would only result in a build-up of excess inventory. This again represents something of a combination of JIT- and MRP-type principles.[2]

All of these developments represent an extension of choice to suit particular manufacturing product/market environments. Under certain conditions, different combinations will be appropriate. Parnaby, for example, suggests a classification into three kinds of products; regular runners, irregular runners and strangers. In an environment which is predominantly making regular or irregular long runs, some form of JIT/MRP is most appropriate. In one which emphasizes high variety – many unfamiliar items – MRP2/JIT, with its sophisticated planning, may be more appropriate.

6.7   THE FUTURE FOR CAPM

Although there is a need to match systems more appropriately to particular company needs, to judge by the predicted market growth figures the overall future for CAPM looks firmly established. This is certainly true for many options such as shop-floor data collection and related data capture systems. Terminals will become more powerful and faster, so that distributed networks will replace hierarchies, making more information available to more users in the form in which they need it. Expert systems will assist in the analysis and some of the decision-making resulting from such information. And finally CIM will mean that not only production management related information but also design data, sales and market information, and other financial aspects will increasingly be brought into the system.

Or will it? CAPM is a good example of a powerful technological trajectory in which need pull and technology push combine to shape a particular and dominant form for the systems which emerge. In circumstances such as these there is a strong self-reinforcing character to the process. Starting from the premise that production management is about information processing, systems have emerged which allow for ever more information to be collected, processed and made available. But in all of this it is important to remember that information is *not* knowledge. We need to look at how people actually *use* such information in their organizations, and how effective a contribution that makes to overall performance.

6.8   INTEGRATION AND SUBSTITUTION

CAPM provides a good example of a technology which can be used in both substitution and integration modes. A typical example of the former would be a stores management or stock control system based on a stand-alone microcomputer, effectively replacing a manual system such as a card index and offering some useful improvements in managing that operation – more accurate record keeping, faster identification of potential stock-outs, and so on. But, ultimately, the scope would be limited to 'doing what we've always done a bit better' and its impact would only be of local significance.

By contrast, an integrated system would offer the chance to share information between different functions in the business, and in the process to enhance considerably the quality and timeliness of information support to decision-making. Benefits here relate to improvements in the entire manufacturing chain rather than in a single area, and emerge in outcomes such as overall lead time reductions and better inventory management at all stages in the manufacturing process. The whole becomes greater than the sum of its individual parts.

A research programme on behalf of the ACME Directorate of the UK Science and Engineering Research Council has been working on the problems of CAPM for some time.[3] One of their first tasks was to take a 'snapshot' of the state of the art of the use of this technology in the UK and, amongst other things, this led to a useful typology of systems according to the level of integration involved, which is reproduced in table 6.2.

Level	Integration status	Definition
0	No CAPM	No CAPM or only beginning installation
1	No integration	Several functions computerized but without regard to integration
2	Partial integration	Several functions linked via common files and coordinated controls
3	Full integration	All CAPM functions linked using common databases
4	Integration of manufacturing systems	CAPM systems designed in conjunction with systems for material conversion, handling and quality and against manufacturing strategy objectives

Table 6.2   Levels of integration in CAPM systems

As we move up the levels of integration with CAPM systems so we might expect the investment costs and the degree of difficulty in achieving successful implementation to increase. This approach forms the basis of one success measure used by systems suppliers to rate their customers as class A, B or C users. According to this scheme, class C users make use of MRP only as an order launching an expediting system. Class B users also use it for production and inventory control, but often rely on other systems such as shortage sheets to effect control. Class A users use MRP to produce and implement the business plan, the implication being that they are able to user the technology to its full potential.

The startling thing to emerge from this crude form of classification is the small number of class A users which exist, even though the basic technology of CAPM (in the form of MRP) has been around for almost 30 years. It is given much sharper emphasis in a number of recent independent studies of the experience of CAPM implementation. For example, the ACME research mentioned above found that at low (= substitution) levels of application, CAPM systems are generally effective, but as the level of integration increases so the effective use falls off.

Of 33 firms studied, 16 claimed to have been successful in their implementation, six were failures and the remainder occupied the middle ground. The researchers conclude that ". . . even advanced CAPM users have difficulty in understanding how best to use the numerous CAPM control variables . . . thus they are not getting full benefit from the CAPM system'.[4]

In another study, New and Myers examined the experience of 167 manufacturing plants using CAPM, as part of a wider study for the British Institute of Management.[5] Of these, at least 20 per cent reported that the pay-offs to date had been zero to low. Another study surveyed 63 companies: almost half of the managers asked (45.5 per cent) reportedly felt that their systems were in some way inadequate.[6] Similar concerns were also expressed in an earlier study for NEDO,[7] while data from the US suggests a similar incidence of problems.[8]

Occasionally problems have been so severe that the system in question has had to be abandoned, as was the case with the UK cycle manufacturer TI–Raleigh, mentioned in chapter 4.[9]

Usually it is not a case of systems failing to perform completely but rather of users failing to obtain the full potential benefits from their systems. Instead of integrated systems being better than the sum of their parts they are in fact less effective than stand-alone systems. For many users, heavy investment in integrated CAPM systems is largely wasted because they are being used in 'dis-integrated' fashion, to support local activities rather than as a powerful integrated information resource to support the whole business.

But why should it still be difficult to achieve successful implementation of CAPM beyond the basic level?

It is important to focus on the location of the main problems. Although CAPM – in common with many other technologies – suffers from technical problems, the major area of difficulty in using CAPM systems effectively appears to arise in the move to higher levels of integration. The ACME study attempted an analysis of the problem and found that there was a high incidence of success amongst users at low levels of integration (as one might expect). The principal challenges here are in providing the necessary skills and ensuring a high quality of data integrity.

The significant finding of the study, however, was that there was also a high incidence of successful implementation at high levels of integration, beyond level 3 in their model. In other words, these users had somehow learned to handle the challenges of integration. The main difficulty was for firms moving from level 2 to level 3; that is, from partial CAPM integration to full integration. Some of the requirements for successfully moving up the integration scale are highlighted in table 6.3.

Level of integration	Technical	Information	Strategic	Functional
0	Self-contained programs – no integration of hardware or software	Local level, high degree of shared understanding – no need for shared information	Local level – clear sense of common aims and purpose	Local level –no need for functional integration
1	Several functions computerized but in local fashion – no need for software or hardware integration between systems	Local orientation and little sharing of information beyond immediate area of application	Local level emphasis – clear sense of local purpose and extent of commitment to system confined to supporting these local aims	Local application – no sharing of information or closer integration between functions
2	Hardware and software integration between functions – common files and coordinated controls	Shared database, and reliance on data input and integrity from more than one source – need for shared view of system and definitions within it	Multifunction impact of system requires shared goals and commitment to a broader organizational purpose	Linking of several functions requiring alternative – more integrated – approach to operation
3	High levels of integration needed within hardware and software	Increasing need for shared information common definitions and new flow patterns	Increasing move towards organizational rather than local goals – effectiveness rather than efficiency as key measure	Increasing merging of functions, with need to break down inter-functional boundaries – emergence of new forms of coordination
4	Full computer-integrated manufacturing – very high levels of integration but also decentralization through networking	CIM depends on use of a common database and information flows geared to networking around this	Depends on high levels of understanding of, and commitment to, strategic goals of organization	Computer-integrated manufacturing will only work in an integrated organization

Table 6.3   Requirements in integrated CAPM

It can be seen that while some of the problems relate to technical issues (especially those surrounding the collection of relevant and timely data) several key organizational and managerial questions emerge. This emphasis is supported by the fact that different users of the same basic

CAPM systems obtain different results, suggesting that the key issues lie less with the technology itself than with the way in which it is being implemented.

6.9   SUCCESS AND FAILURE IN CAPM IMPLEMENTATION

There is now a broadly based range of experience regarding success and failure factors in CAPM implementation, representing the different perspectives of users, suppliers, consultants and researchers. For example, a recent study suggests that the success rate reported amongst professionals implementing MRP2 is less than 50 per cent.[10] Reasons for this include:

- a lack of understanding of MRP2, at the level of its being a management philosophy rather than simply a tool for inventory control

- a lack of commitment from senior management

- a lack of education – as one consultant put it, 'too little, too late, wrong people'

- a lack of communication – MRP2 is an integrated system and requires a similarly integrated communication pattern within the organization if it is to work effectively

- the need to change the basis of measurement of performance to something which is less concerned with narrow efficiency and more with achieving broader organizational objectives – effectiveness rather than efficiency

Hughes et al. identified three main causes of CAPM failure:[11]

- the system did not meet requirements (even if it met the actual specification) – this reflects weaknesses in the planning and selection process

- new requirements emerged later which the system could not meet – this again reflects a lack of foresight and strategic thinking in the selection process;

- poor implementation and project management

In addition, Hughes et al. make the point that many failures cannot be attributed to these 'traditional' causes but are, instead, symptomatic of a failure on the part of the organization and its systems to adapt to a rapidly changing environment. Even if their CAPM systems are currently performing well, the critical question is whether the combination of organization and system has the necessary flexibility and adaptability to respond to unexpcted future changes.

A significant point made in several studies is the need for simplification and reorganization as a precursor to CAPM implementation. In many cases failure is due to firms trying to computerize systems which are already too complex and cumbersome, the contribution of MRP2 being to worsen the complexity. By contrast, one of the main advantages of a 'just-in-time' approach is that it explicitly requires a simplification and a rethink of production and inventory management.

Concern has also been expressed about the mismatch between MRP2 and the dynamic business environment.[12] MRP2 is founded on many of the old views of production and:

- concentrates on the long term, with a lack of flexibility and stability

- is traditionally about high volumes

- is based around traditional accounting practices

- is based around functional machine layout for batch operation

- is based on strict division of labour

- assumes limited workforce responsibility

- ignores issues of quality

- focuses on resources and does not take capacity constraints into account

In other words, it emphasizes many *dis*-integrating practices while offering an integrated system. It has been suggested that in order to make such CAPM systems more effective greater attention must be paid to challenging and changing some of these factors – again arguing for the use of JIT/MRP2 hybrids which build upon the idea of a fresh approach to production management.

Not all CAPM systems fail, but it is significant to see further emphasis on the managerial and organizational aspects as key influences on obtaining *successful* performance. For example, in answer to the question, 'What makes for success?', experienced users and consultants identified the following:[13]

- Not just commitment from senior management but an understanding of the major business impacts of such a system, the responsibilities they will be required to undertake and the implementation resources required, in terms of money, time and personnel.

- Middle management are crucial to success: 'If MRP2 does not operate effectively in every department, it will never operate effectively at the fully integrated company level.'

- Be prepared to change existing practices.

- Education is central to the process.

- Select team members with care.

- Engender understanding, justification and commitment by jointly developing and agreeing a formal blueprint. This represents a shared vision of the system which all those involved in its preparation should formally sign off as an expression of their commitment to it.

- Once the blueprint is ready, then begin to involve the rest of the company. In order to transfer ownership, use presentations which mirror or cascade the same strategy and vision-building process down through the organization.

Successful users lend support to this view, as the following quotes indicate:

> Our experience is that you need the tools but you can do abysmally with them unless you have the support of senior managers and companywide commitment.

> Switching the system on is only the first step – obtaining the best results takes much longer.

This second company introduced MRP2 in 1986 but needed concerted management action and help from external consultants to get the best out of their system. They emphasize training as a key factor, not just at the level of operating the system but at all levels: even with an average of one week per manager and users plus more informal training they felt this was 'only adequate' and that more was needed. A second key factor was commitment:

> Credibility in the system must be established quickly and this can only be achieved when a high level of visible management commitment exists and problems are dealt with swiftly, competently and with patience.

Finally, a clear strategic view of where the system is taking the organization is needed:

> 'Don't just try and computerise what's already there!'

Mention should also be made of work carried out over many years by a team led by Professor Enid Mumford at Manchester Business School, into some of the human factors and organizational dimensions of implementing computer systems. This has led to the development of an approach they call ETHICS (Effective Technical and Human Implementation of Computer Systems).[14] This involves a sequential innovation process but also emphasizes some critical factors that are not always considered in the implementation of computer systems: user design of systems (or at least, their part in those systems); continuous monitoring and design input during the changeover process; changing the design of jobs and of organizational structures and procedures in parallel with the technical change; and a continuing commitment to building and implementing a shared vision of the project and its strategic objectives. It is interesting to note the striking similarities between these factors and those presented above by both practitioners and users.

Significantly, if we distil the main themes in all of these success and failure factors, as in table 6.4, they mirror each other closely. In other words, there are certain kinds of behaviour which actively facilitate success and the absence of which heralds failure.

Success is associated with . . .	Failure is associated with . . .
Top management commitment at all stages of the project	Lack of commitment
Clear strategic vision, communicated throughout the organization	Lack of clear strategy and/or its effective communication to the rest of the organization
Shared views of project aims and implementation approach	Lack of shared view and unresolved conflicts regarding design and implementation
Multifunction project teams with multifunction perspective	Single-function teams, unilateral perspective
Effective conflict resolution within team	Unresolved conflicts over key implementation issues
Extensive user education to give understanding of broader implications and purpose of system	Minimal training for operation
User involvement in system design (of hardware/software, jobs, structures roles, and so on)	Unilateral design, organization expected to adapt to systems rather than change system
Close involvement with suppliers	Minimal involvement
Readiness to re-examine and change existing procedures	Attempt to computerize what is already there
Performance measures reflect broader organizational effectiveness	Performance measures narrowly defined and related to efficiency at local level
Flexibility in design and continuous monitoring to adapt to unexpected changes	Inflexibility in system in response to unexpected changes in environment

Sources: Mumford,[7] Hughes et al.,[8] Gillingwater,[9] Voss[10] and Trade press[11]

Table 6.4   Success and failure in CAPM implementation

6.10   ORGANIZATIONAL LEARNING AND CAPM

The move to CAPM can be seen as following a learning curve, the early stages of which are relatively simple, and along which progress is usually rapid. But as the focus shifts from simply acquiring skills to use new tools in substitution mode and towards changing the way in which the organization operates, so the challenge becomes much more difficult. One way of interpreting this is through the idea of organizational learning and adaptation. For as long as the challenges remain at a manageable level, within the organization's existing capacity to change itself, the innovation will succeed. But as the demands posed by the technology require higher levels of adaptation, so the organization moves beyond its traditional experience and needs to learn and develop new approaches.

For example, the principles of bureaucratic organization (which still dominate most structures within manufacturing) operate in ways which may not necessarily assist the use of highly integrated information systems. In particular:

1   They emphasize fragmented structures, which means that knowledge and information about what is going on, where the organization is heading, and so on, tends to be fragmented and incomplete, or filtered through a particular local 'way of seeing'. Such fragmentation also inhibits the free flow of information – even if the technical systems (such as computer networks) facilitate this. The result is that different parts of the organization often work with differing information and differing pictures of the whole situation and – as a result – they often pursue local rather than total organizational goals. Such distinctions inevitably fuel interdepartmental and other forms of political behaviour.

2   Bureaucratic systems are organized on the basis of clearly defined roles and responsibilities, and tend to reward people for maintaining their place within this system – not for challenging it, or the policies and procedures whereby it operates.

3   Bureaucratic systems operate on the basis of accountability whereby success is rewarded and failure punished. Consequently, behaviour is often geared towards protecting the individual or unit, and this may lead to various forms of deception, or obscuring of the relevant information so as to hide negative performance or present it in a good light. This means that information is often incomplete or inaccurate and also creates a climate of defensiveness. In such circumstances it is often difficult to tolerate high levels of uncertainty: instead, there is an attempt to simplify complex problems or to treat symptoms rather than explore their deeper root causes. Equally, simple solutions are sought –

technological 'fixes', for example – in the mistaken belief that there is a single, 'right' answer or cure.

When the move is from level 0 to level 1 – the discrete substitution of manual systems within a local area by computer-based systems (often a stand-alone PC with no attempt at networking) – the challenge posed to the organizational base is relatively small and confined to one area, so the influence on the speed of progress along the learning curve is slight. In other words, it lies within the organization's existing capacity to manage learning. But when the transition involves a quantum leap to integrated systems, which have implications for different functional areas, which may upset the balance of power or the flow of information, which may be the subject of different views about goals, and so on – then the challenge is much greater and lies outside the organization's capacity to learn – it requires some adaptation and development.

That some firms have moved beyond this suggests that they have already managed to acquire this additional learning, and it correlates well with the finding that the move from level 2 to 3 is much easier. This suggests that experience is important, and that successful level 2 firms have found ways of handling the organizational questions raised by integrated systems. There is some empirical evidence which might support this hypothesis. First, if we look at some of the common symptoms of level 2–3 systems (such as MRP2) which have not fully succeeded, it becomes possible to interpret them in terms of this learning model. Some examples are given in table 6.5.

Symptom	Organizational characteristic
Poor data integrity	Lack of openness and information sharing, with a unilateral rather than a global view – 'information is power, so keep it close to your chest'
Integrated system (level 2+) used in a non-integrated way – for example, only some modules of MRP2 used and confined to local areas only	Lack of shared vision of system or goals – unilateral design, exclusion of some key user groups
Lack of long-term flexibility in the face of a changing environment	Inability to revise goals and adapt to meet new challenges
Continuing dependence on manual systems in parallel	Lack of commitment and trust in the new system and in the information provided by other functions
Interdepartmental conflicts	Unilateral views in design and lack of effective conflict resolution in implementation
Inadequate skills to support implementation	Lack of training, itself a symptom of lack of communication and involvement

Table 6.5   Organizational learning and CAPM: symptoms

By the same token, our evidence about success factors suggests a number of features which are associated with alternative approaches. Again, some examples are given in table 6.6.

Success factor	Organizational characteristic
Participative design	Generate commitment and high-quality solutions through participation of all those with valid information to contribute
Multifunction project teams	Shared perspective on system
Top management commitment	Shared strategic vision, which is effectively communicated downwards throughout the organization
Readiness to re-examine and change existing procedures	Monitoring and adopting on a continuous basis
Close working with system suppliers	Making use of all valid information inputs
Cross-functional cooperation	Effective conflict resolution as a part of building a shared vision and agreeing a common blueprint
Extensive and widespread education and training	Building and communicating a shared understanding of the project and its relevance to the business
Continuous monitoring and design input throughout project and on all aspects of the system – software, hardware, jobs, structures and so on	Monitoring and adapting on a continous basis

Table 6.6   Organizational learning and CAPM: success factors

Most of the best-practice prescriptions can be reinterpreted in this light. Participation is about securing shared ownership, and communication is about maintaining a shared understanding. Strategic integration is about ensuring that the views about the needs for and goals of a system are aligned and congruent. Training helps to develop operating skills, and also to build up a shared picture of what the system is about. In other words, the organization needs to adapt and develop in parallel with the technology being implemented. But, beyond this, it needs to acquire new ways of monitoring its activities and changing its procedures on a continuing basis. This move towards becoming a 'learning organization' is critical to success, not just in CAPM but in the context of all the integrated technologies discussed in this book. We will return to the theme of organizational learning in chapter 13.

6.11   SUMMARY

In this chapter we have attempted to highlight a number of key points about CAPM. First, it represents a powerful technological trajectory which has dominated production management thinking over the past 30 years. CAPM systems now offer a very wide range of choice and availability, and many of the earlier technical problems – of access, availability of terminals, software friendliness, and cost – have been solved.

The main problems in successful implementation relate to the level of integration and are primarily organizational/managerial in nature. At low levels the task is to ensure skills, to train for data integrity and to obtain commitment and acceptance of change. Such change is relatively minor because it is largely substitution innovation.

In moving to higher levels of integration the need is for more substantial organizational change. It includes finding ways of integrating different functions more closely so that they share information, of ensuring a common vision of what the system is and why it is being implemented, and of building commitment and ownership of the system. Processes include extensive training, not just for specific skills but to increase awareness and understanding of the strategic role of the system, participative design and implementation of systems through which ownership and commitment can be built up, the continuance of top-level commitment and support for a multi-functional project team, and effective communication throughout the process.

Such a radical change in the behaviour of the organization is not achieved painlessly: evidence suggests that it needs considerable investment of time and money and often requires external agents, consultants, and so on to facilitate it. Not all firms succeed, but it does appear that this kind of approach, based on on a more open and co-operative commitment to a clearly defined common goal, offers a recipe for successful implementation not only of CAPM but also of other integrating technologies.

Notes

1   D. Corke, *A Guide to CAPM* (Institution of Production Engineers, London, 1985).
2   For example, see E. Goldratt and J. Cox, *The Goal*, (Creative Output, New York, 1984).
3   G. Waterlow and J. Monniot, 'The state of the art in CAPM in the UK', *International Journal of Operations and Production Management*, 7 (1) (1986).
4   Waterlow and Monniot, 'The state of the art'.
5   C. New and L. Myers, *Manufacturing Operations in the UK*, (British Institute of Management/Cranfield Institute of Technology, 1985).
6   D. Gillingwater, *Attitudes to the Use of Computerised Information Systems for Production Management in Manufacturing Industry* (Department of Transport Technology, University of Loughborough, August 1987).
7   NEDO, *Computers in Production Control*, (National Economic Development Office, London, 1984).

8   T. Callarman, 'A model for MRP implementation', in *Managing Advanced Manufacturing Technology*, ed. C. Voss (IFS Publications, Kempston, 1986).

9   *Computer Weekly*, 23 February 1989.

10 Cited in *Automation* magazine, March 1990.

11 D. Hughes and S. Childe, Report to the ACME Directorate, SouthWest Polytechnic, Plymouth,1990.

12 A Yankee Group report, cited in *Industrial Computing*, January, 1989.

13 *Works Management*, October 1987.

14 E. Mumford, *Designing Human Systems*, (Manchester Business School, 1982).

7 Design

7.1 INTRODUCTION

Design is a key feature of many manufacturing businesses – and its importance is growing. As emphasis moves towards non-price factors, so design becomes a major determinant of competitiveness, with a marked shift in company strategies regarding design and product innovation, stressing novel features, frequent changes, greater response to customer – specific requirements and more rapid new product development cycles.

All this demands a new attack on the design process, not only employing new tools such as computer-aided design (CAD) but also new approaches to managing design as a more integrated theme within manufacturing. As companies (particularly in Japan) begin to achieve major improvements on the factory floor, so their attention is now turning to the design and product innovation process, aiming to reduce the 'time to market'[1] For example, Hayes et al.[2] cite the case of the Honda company which, through systematic improvement of its total design process, has now reduced the time taken to launch a new model car to less than three years, while the time needed to implement complex engine modifications has now been cut to eight weeks.

This increasing importance of design is widely recognized. In the UK, for example, a series of studies in the 1970s drew attention to the relatively weak state of UK design and, more recently, the Design Council has taken a strong lead, with enthusiastic support from no less than ex-Prime Minister Margaret Thatcher herself.[3] One practical outcome of this has been the proliferation of a range of design support schemes aimed specifically at improving design capability and practice within the manufacturing sector, particularly for the smaller firm. Similar schemes operate in other European countries, including the Federal Republic of Germany. Much of this effort has been devoted to the promotion of CAD technology, which can offer considerable opportunities for improving design activity in even the smallest firm.

7.2 NEED PULL

The activity of design involves several things: it takes ideas for new or improved products, customer specifications, safety standards and other regulatory information, information about process capability, materials and so on, and converts them into a drawing containing detailed information about a product. This information also provides inputs to other activities in manufacturing, such as:

- to the coordination area, where purchase orders, works orders, and so on are raised to ensure that materials are available, and to production scheduling to ensure that capacity is available, and to quality control to keep track of the final product conforming to the specifications, and so on.

- to the production area itself, with details of how the machines need to be set, fixtures built, moulds and dies constructed, and so on.

- design is also involved in the development of new processes and in the improvement of existing ones – for example, in laying out pipework in an oil refinery for optimum use of energy, or in routing materials and sub-assemblies through a plant to ensure optimum plant utilization

Such a process involves several skills but, for convenience, we can break them down into two groups:

- *design* skills, which are essentially conceptual, synthesizing a design from this mixed information input

- *draughting* skills, which are essentially the physical skills of drawing accurately

Since even a simple product requires extensive information to support its manufacture, it is not surprising that most firms which have their own design department, traditionally recognized by rooms full of drawing boards, each with a draughtsman painstakingly detailing his particular drawing. Even those firms which do not support much in-house research and development (R & D) will usually have some form of design capacity to translate customer orders into production drawings and/or to maintain and update their in-house tooling and equipment.

While in the past design could be considered the concern of a small number of firms, the requirement to offer variety in product range and more frequent model change is challenging firms across the manufacturing spectrum to increase their design capacity, as one more way of becoming more flexible and responsive. Design can also contribute to improving competitiveness by helping to save materials, to simplify production processes (by reducing the number of operations or the time taken to carry them out), to improve fixturing, and so on.

Some examples of design activities in different industries include the following:

- in clothing – patterns, styles, tailoring, regular updates to keep up with the fashion market, cutting to optimize material usage

- in engineering – drawings and blueprints for parts and fixturing, engineering calculations, structural analyses, responding to customer-specific requests (especially in the subcontracting industry)

- in textiles – patterns, label design

- in furniture – drawings, layout plans for integrated suites or rooms (such as kitchen or bathroom planning)

- in process industries – pipework layout, flowcharting

- in electronics – printed circuit board (PCB) layout, integrated circuit design (mask-making for lithography)

- in food processing – packaging design, process flowcharting

Despite widespread experience in different applications, delivering effective design is not a simple process. A number of problems emerge with traditional manual systems, including:

1 *Lead time* (to respond to customer order or introduce a new product). The speed with which a drawing can be prepared depends on manual draughting skills. Many activities, such as creating borders, lettering, repeating patterns, and so on, are laborious and add considerably to the overall design time while also increasing the risk of error. Given that for many businesses, especially in the subcontracting field, delivery and rapid response to customer needs are critical factors determining competitiveness, any delays in the design process will be a major problem. In addition, much competitive success is based on the ability to introduce new products rapidly; and with product life cycles for some items – for example, computers or consumer electronics – often measured in months, such time-to-market advantages are critical.

2 *Errors.* During the process of draughting, errors may creep in, especially where the design involves a large number of similar but slightly different elements. To some extent there is a trade-off between working faster at design so as to be able to respond rapidly to the marketplace and the emergence of errors as a consequence of less careful work.

3 *Lack of connection to manufacturing.* In many cases design takes place as a separate activity from manufacturing. One consequence of this is that errors which have not been picked up in the design process will only emerge when the product in question is manufactured for the first time. The whole process then needs to start again, with the drawings being returned to the design area for

error correction, a process which clearly adds to the overall design time.

4   *Lack of simultaneous working.* For complex designs, which involve different groups working on different but related drawings, there is the major difficulty of ensuring that everyone is working on the same version of the overall design. If we consider the case of a complex assembly such as a motor car, then it is possible to have many thousands of different drawings being worked on at any moment – detailing the engine and power train, the doors, windows, steering, braking system, and so on. Many of these are interrelated, so that a change in one will affect a design that someone else is working on. Unless this information can be communicated quickly there is a real risk that incompatibilities will emerge – and that parts will not fit together.

5   *Storage and retrieval.* For many businesses, design activity involves extensive re-use of designs, or minor variations on basic designs. Products and parts going back over many years, may also be kept active, so that there is a need to retain designs (in order to be able to produce a part or product) for what is often a very large number of potential orders. Managing this information in terms of storage and retrieval is an increasing problem with manual systems, especially in terms of the physical space in which to store so many drawings.

As the demand for greater design activity increases and spreads across the manufacturing spectrum, so the need for new tools to help deal with these problems has grown and led to the emergence of CAD.

7.3   TECHNOLOGY PUSH: THE EMERGENCE OF COMPUTER-AIDED DESIGN TECHNOLOGY

Early days

The principles behind CAD are not new. Interactive graphics was first used in the US for the SAGE early warning radar programme. As early as 1959, a group working for the USAF at MIT produced a specification for a basic CAD system using a mainframe computer. Credit is usually given to Sutherland for his 1963 work on a project known as 'Sketchpad', which aimed to create what by today's standards was a crude 'electronic pencil' that would provide an aid to basic draughting activities and thus increase speed and productivity in the design and manufacture of aircraft components.

The basic requirements for such a system have not changed dramatically since then and are:

- some form of electronic drawing board which permits design information to be converted into electronic data

- some form of information processing which accomplishes the conversion and stores and manipulates the resulting images in electronic form

- some form of display device (such as a visual display unit) to show the resulting design in its electronic form

- some form of hard copy producer (such as a printer) to produce a permanent physical record of the design

Although relatively simple in concept, the massive information-processing problems (in terms of power, speed, display technology, and so on) all meant that early development and application was limited by high costs and related barriers to a few sectors of industry. The use of these first-generation tools was confined to industries with a major design requirement, such as:

- aerospace – the technology originally emerged as a result of US military funding for applications in this industry (for example, by 1965 IBM, McDonnell and Boeing were experimenting with versions of CAD/CAM, linking their CAD systems to NC machinery)

- electronics, where by the late 1960s new design tools were needed to support increasingly high levels of integration in circuit manufacture

- cartography

- construction (for complex projects)

- the automobile industry, where General Motors had begun work on CAD back in 1959 and was routinely using interactive graphics in its design process by 1963[4] – Ford followed in 1964 and other manufacturers joined by the end of the decade

Each of these had different needs which increasingly began to shape the development of CAD. Once again, we can see the emergence of an innovation cycle in which the emerging technology opened up opportunities which were than applied to particular sectors. In turn, their increasingly complicated requirements fuelled the development of the next generation.[5]

The turnkey era

A major breakthrough for the industry came in the 1970s with, as Arnold[6] points out, the emergence of several key inventions. These were:

- the minicomputer, which offered considerably greater computing power at lower cost than its mainframe predecessors, and opened up the possibility of using dedicated computers to support CAD

- the Tektronix storage tube, which solved many of the problems associated with display

- virtual memory concepts, which helped ease the problem of access to processing and display

- new programming tools, especially the use of structured programming approaches

The effect of these was to open up the CAD industry to a wider range of potential applications, and to bring those which had been developed to a larger number of firms. At the start of the 1970s there were around 10–15 suppliers of CAD, (mostly US-based) but by the end of the decade there were over 100, distributed throughout the world. This period also saw growth in specialist applications aimed at solving particular types of design problem; for example, specialist systems for layout in electronics, or for pipework design and layout in the chemical and process industries.

Amongst the key features of this diversification during the 1970s were moves towards multidimensional representation. Early CAD systems had been based on two-dimensional representation, or at best 'wire frame' representations of three dimensions (sometimes termed '$2^{1/2}$-D'). These essentially reproduced conventional drawing board representations and were adequate for many applications where the main task of CAD was to improve conventional drawing office practice. But it also became clear that other more specialized needs could be catered for; one example is the use of multilayer 2-D models to facilitate layouts of electronic circuits in the mask-making process for VLSI chips.

A second area which required considerably greater processing power and more complex software was three-dimensional surface modelling, in which the wire frame was replaced by a 3-D visual representation. The problem with wire-frame models (which look as if the shape they represent has been constructed out of thin wires) is often one of image perception; for example, it is often hard to convey depth in such images. To appreciate the shape effectively there is a need to introduce 'faces' between the different wire points and at least give the impression of solidity. Such modelling opened up new possibilities in aesthetics, and also in a variety of simulation and modelling activities. For example, it became possible to 'see' what a final version of a new product might look like and how it might compare with other options.

A later development, which required even more complex software and processing power was the development of 3-D solid modelling

(sometimes called volumetric or geometric modelling). In surface modelling the picture looked three-dimensional but, to the computer, was in
fact a wire frame on which a surface skin had been stretched – it was still
a representation, not a complete model. By contrast, 3-D solid modelling
involves a full model built up from individual elements, which is a repli-
ca of the object in question, held in the computer memory. As one writer
put it, 'a 3-D solid model has an identity – it knows what it is like from
the inside, not just the outside'. The considerable advantage of this kind
of CAD model is that it can be used to simulate the behaviour of the real
thing: for example, the response to different environmental conditions or
forces.

Finite element analysis (FE) is another tool which allows the construction of such solid models and which permits powerful simulation of
behaviour of structures. In essence, it involves creating a computer
model by breaking the structure down into tiny single elements, the
behaviour of which in response to various forces and conditions can be
accurately predicted via mathematical models. The behaviour of the total
structure is then simulated by summing the behaviours of the finite
elements: a set of complex calculations which could only be carried out
fast enough through the use of extensive computer power and advanced
software techniques. FE analysis is extensively used in civil engineering
projects and in examining the effects of physical forces on products; for
example, of the dynamics of motor vehicles or the performance of
pistons within engines.

Simulation is another powerful feature of CAD systems, using a
variety of software techniques and mathematical models. Basically, the
principle behind simulation is that it is cheaper, faster and easier to
explore the behaviour of products and systems before they are made or
used, through simulation in a CAD system. For example, safety legisla-
tion requires that motor vehicles are subjected to crash tests in order to
ensure pasenger protection. Each test costs around £100 000, and while it
yields a wealth of information it is a costly exercise. Using FE analysis
and other simulation techniques it is possible to carry out hundreds of
crash simulations and experiments, and thus to extend the range of
information to improve the final product design.

Other applications include ergonomics (for example, simulating how
people sit in a car and how easy/difficult it is for them to access controls,
and so on) and dynamics (how particular products behave when in
operation). Such simulation does not only apply to products: it can also
be used to explore new manufacturing systems or layouts. In applications
of this kind, which can be likened to 'test driving' a new facility before
full operation, savings can be 'typically up to 10 times the cost of the
simulation', obtained through optimal use of materials and resources.[7]

For example, the layout of a flexible manufacturing system can be optimized (in terms of positioning of machine tools and operation of handling and transport systems) by simulation of its performance under different factory order loads.

For many organizations a key design problem is the storage and retrieval of information relating to wide and old product ranges which are still active and for which orders may be anticipated. CAD systems permit the use of computer storage to hold electronic information and the use of database management techniques to radically improve the speed and efficiency of access. A further advantage of such systems is that where a new order is only slightly different to a part already held in the library, the library drawing can be modified quickly to create the new drawing – thus cutting the overall response time.

One final system which first emerged during this period was CAD/CAM, computer-aided design and manufacturing (sometimes called CAE, computer-aided engineering). This involves extending the link beyond the 'pure' design area into the later stages of the manufacturing process. Since the information about a part or product is generated and held electronically in a CAD system, and the operation of many pieces of production equipment is now under electronic control, the potential exists to use the computer to generate information not only for the product itself but also the control of the process for its manufacture. The benefits of such integration are enormous – reductions in lead times, improvements in quality, savings in materials, and so on – but the challenges posed by these systems are also greater than for systems aimed only at design.

Characteristic of the systems sold during the 1970s was their 'turnkey' nature: that is, they were often large systems with a limited number of screens, but which offered entry into the whole range of these applications. This extended the application of CAD considerably, but the relatively high cost and complexity of such turnkey installations tended to exclude the smaller organization.

The emergence of low-cost CAD

A second surge of growth occurred in the early 1980s with the emergence of the PC-standard computer, orginally based on the IBM PC and compatible machines. This effectively brought the processing power and display capability of the original 1960s experiments within the reach of a very wide range of potential users. In particular, it became clear that for many basic draughting tasks for which 2-D CAD and relatively simple software were required, it would be possible to offer low-cost solutions which could be packaged to make them much more accessible and friendly to the user. Earlier generations of CAD had suffered from being seen as requiring expensive training investments in addition to the hardware and software costs. Access was limited because of the high costs of seats at a networked system, and this effectively excluded smaller businesses unless they were able to use a bureau.

The availability of low-cost CAD meant that it was possible to operate CAD in almost any firm and to distribute stand-alone CAD systems widely. Not surprisingly, growth in this market sector was rapid and by the mid-1980s the suppliers of such systems (such as Autodesk, producers of 'Autocad') were the market leaders (measured in terms of volume, not value of systems sold). One consequence of this was the rapid proliferation of suppliers writing software for CAD. The synergy between a rapidly growing user base and new entrants to the supply side (estimates suggests that by 1985 there were over 1000 CAD suppliers worldwide) meant that the rate of innovation was high, especially at the lower-cost PC-based end of the market. Thus, features which had been confined to larger systems – such as '$2^{1/2}$-D' or even 3-D modelling, or simple CAD/CAM – began to be made available on PC systems. Inevitably, this has led to considerable segmentation of the market and to the proliferation of choice.

Future trends

In future this range of choice is likely to be extended further. With increases in the power of PC-type computers and decreases in the costs of workstations, the potential exists for an individual to have access to a powerful CAD/CAM system based either on a stand-alone workstation or linked into a wider network. Many of the market leaders for large turnkey systems have now produced versions of their software and systems which can be run on stand-alone machines and which are also upwardly compatible with their larger systems – effectively allowing for users to grow with their CAD systems. At the same time, developments within the area of artifical intelligence and advanced software techniques are opening up considerable opportunities in the area of 'intelligent CAD', where many design activities in addition to basic draughting can be undertaken automatically.

An example of this trend is a system offered by a small US software house called ICAD. Here an attempt has been made to move CAD into a 'fourth generation' (the previous three being 2-D draughting, 3-D models and solid modelling), using rule-based systems to automate designers' decisions as well as their pen movements. One company in which this approach has been used (an engineering firm producing heat exchangers to customer-specific requirements), was faced with the problem that each exchanger had thousands of parts, defined on over 100 000 drawings. The manufacturer used 7000 design rules, which have now been built into the system: these can now be used to modify designs so that the firm can draw up as many as 30 proposals a day using only five design engineers and the ICAD system.[8]

7.4   DIFFUSION OF CAD

Not surprisingly, the market has segmented in response to the emergence of such wide choice. In very broad terms there is now a split between the use of CAD as a draughting aid – the equivalent of a word processor for the draughtsman, which permits improvements in productivity but is really little more than an electronic drawing board – and its use in full *design* activity. The pattern has been further complicated by the emergence of the PC system which now dominates the field. Although early applications of such PC-based systems were all in the field of computer-aided draughting, the increasing power of such microcomputers (through the use of 32-bit microprocessors, advanced operating systems which support networking, and so on) has meant that whereas they were originally seen as giving '70 per cent of the power of a workstation for 20 per cent of its cost' they are now becoming direct competitors.

The overall market continues to grow: for example, the 1988 estimate by Dataquest was $11 billion worldwide, with the US accounting for about $4 billion of this and Europe for just under $2.5 billion. According to this survey, growth in the US has levelled out, but in Europe continues to accelerate; this market is estimated to be growing at around 9 per cent per year, and is expected to account for about 30 per cent of the total world market by 1991.

Despite the size of this market and the respectable growth rates by revenue (around 9 per cent in 1987), there is a slowdown from the heady days of the early 1980s, when growth in revenue was running at over 50 per cent per year. Much of this can be explained by the shift in systems, from mini-computer systems to PC-based systems costing far less but offering most of the required features. In 1985, for example, over 80 per cent of European sales were for minisystems, whereas by 1987 85 per cent of the CAD seats sold were provided by PC- and workstation-based systems. The growth in market share by value is less dramatic but still significant: 1988 estimates by Frost and Sullivan suggest that PC-based systems accounted for 16 per cent, workstations for 39 per cent and minicomputers and mainframes for the remaining 45 per cent.

This pattern is also reflected in the supply side. Whereas the industry was dominated in the late 1970s by the major US suppliers – Computervision, Calma and so on – this pattern has shifted with the rise (measured in terms of both units and revenue) of the PC-based producers such as Autodesk. In 1984 the top five suppliers were Computervision, IBM,

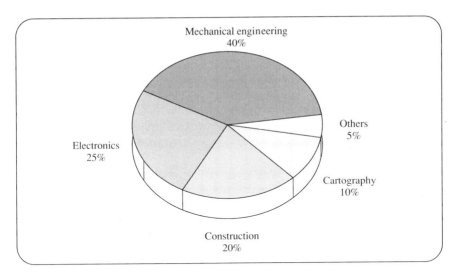

Figure 7.1   The application of CAD

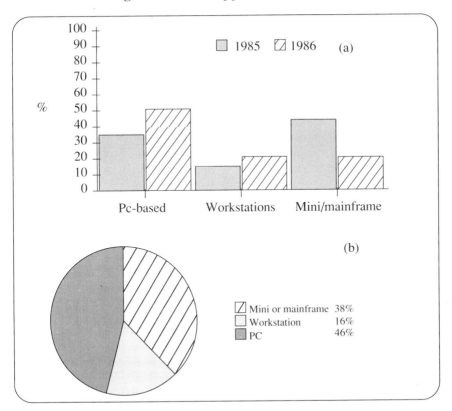

Figure 7.2   Diffusion of CAD (a) in Europe and (b) in the UK by system type, 1988

Intergraph, Calma and McAuto, but by 1987 the top three were Computervision (12.9 per cent of the world market), IBM (11.7 per cent) and Autodesk (10.5 per cent).

In terms of diffusion by application sector, the dominant users remain in electronics and mechanical engineering. The 1987 breakdown for the European market is shown in figure 7.1. In the UK engineering industry, estimates suggest that around 60 per cent of potential users have now adopted the technology, with over 96 per cent of those employing more than 500 employees.[9] This compares with other UK survey data which gives a 1987 figure for mechanical engineering of 50 per cent and of 33 per cent for all industrial sectors.[10] US data suggests a similar pattern for users in engineering and related sectors (SIC headings 34–38), with CAD being used by just under 40 per cent of firms, CAD/CAM by nearly 17 per cent and CAD output used in procurement by just under 10 per cent.[11] As with the UK engineering sector, firms employing over 500 people had a very high rate of use (82.6 per cent).

The diffusion by type of system is shown for the electronics and engineering sectors in the UK in figures 7.2 and 7.3, highlighting the extensive growth in relatively simple systems for basic draughting applications.[12]

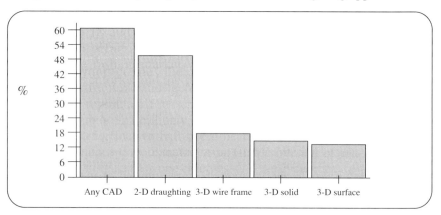

Figure 7.3   Diffusion of CAD by system type in UK mechanical engineering

7.5   EXPERIENCE WITH CAD

Motives

In a recent survey,[13] motives which firms gave for implementing CAD included:

- to reduce lead times in responding to customer orders
- to reduce the time taken to make modifications

- to reduce new product development lead times

- to improve utilization of raw materials

- to improve the quality and accuracy of design

- to simulate and investigate alternative options

- to improve tendering and their overall image to customers

- to reduce errors in complex designs

- to improve drawing office productivity

Arnold and Senker carried out a detailed study of 34 user firms in 1982, and give the following breakdown of motives for investment in their sample, drawn from the aerospace, motor vehicle, electronics and mechanical engineering sectors:

- design flexibility/complexity – 27 per cent

- lead time reduction/viability – 27 per cent

- skill shortage/need to reduce dependence on contract draughtsmen – 18 per cent

- experimental/unclear – 28 per cent

Perhaps the most important point to emerge here is that CAD is clearly not a technology introduced purely for labour-saving purposes in the drawing office, but a product of a much wider set of objectives. However, it is still most commonly *justified* on the basis of improvements in design/draughting labour productivity. In their study Senker and Arnold found that:

> . . . in terms of competitiveness, the most critical advantage was probably reduction in lead times. But in terms of obtaining top management approval for the expenditure necessary to install systems, a vital factor was the ability to demonstrate the likelihood of productivity improvements.

Benefits

What are the benefits in practice? The basic idea of a CAD system is that the draughting activity normally carried out on a drawing board is replaced by an electronic system. This offers benefits which include:

- increased productivity at the draughting stage following from higher drawing speeds and easier correction of mistakes (similar to word processing)

- repeat designs, and designs which only involve minor modifications of a standard design, can easily and quickly be effected by calling them up from a library of drawings and simply changing them on screen

- speed of response in producing large numbers of drawings – especially to support major items of capital equipment – can be increased because much of the work does not have to be started from scratch

- for many draughting activities, such as drawing borders, lettering, scaling, and so on – the system can eliminate the chores by using standard routines which are stored in the memory

- maintaining a library of drawings can save valuable time in searching and redrawing – for example, a typical engineering firm may have 50 000 drawings to support its product range, of which only 20 per cent will be drawings of the product itself, while the remainder refer to fixtures, tooling, jigs, and so on

- as a result of increased speed the overall lead time to develop a new product or modify an existing one to suit a customer's specifications becomes significantly reduced

Thus users of basic systems report benefits of anything up to 300 per cent improvements in draughtsman productivity. Kaplinsky, quoting US and UK data, reports productivity gains of between 2:1 and 27:1, depending on the area of application. In his own studies he found some cases of 200:1 improvements, but he makes the point that much of this benefit is derived from different ways of organizing the design process. For example, the significant improvements in secondary design productivity can be attributed to the storage of engineering information on a database rather than in filing cabinets. He comments, '. . . these productivity ratios depend critically on a change in the organisation of the design process and particularly in a change in the nature of work'.[14]

Arnold and Senker make a similar point, arguing that most turnkey CAD systems came from US suppliers. US drawing offices tend to involve a large number of relatively junior draughtsmen, in contrast to European approaches in which more integrated design teams are used, with a greater involvement in the overall design process. Thus claims for productivity gains based on US experience may not be realistic for the European market.[15]

The whole issue of productivity gains is further complicated by the difficulties involved in measuring and comparing. For example, a simple increase in the number of drawings produced is not a reliable indicator, since this does not take into account the quality or complexity of the designs involved. Then again, there is the question of time spent away from the actual drawing board. Design involves interaction, discussion and individual reflection; even a tracer will spend some time away from the board, and a design engineer may spend up to 80 per cent of his time in non-drawing activity. It has been suggested that an 'average' draughtsman may only spend half of his or her time actually drawing.

Despite the advantages of increased productivity, for many users the main advantage is in the reduction of lead times. This comes about through the overall speed increases in the design process, and also in the use of information previously stored on a database, so that for similar products little new design work needs to be carried out. Kaplinsky cites a particularly dramatic case illustrating the benefits in lead time reduction: this involves a small engineering company that was under threat of closure if a new product could not be launched within 12 months. The traditional design process would have required 12 man–years of draughting, but the situation was exacerbated by a shortage of draughtsmen, so that the minimum lead time for developing the product would have been 18 months. However, using three draughtsmen on shifts around a two-terminal CAD system, the firm were able to generate the 8000 drawings (as opposed to 12 000 for the old manual system) needed to meet the deadline.[16]

In another case an engineering company had a lead time of around 18 weeks for special orders, made up of six weeks waiting for draughtsmen, six weeks in the drawing office and six weeks in production. After installing CAD it was possible to reduce the design cycle to two weeks, giving a total lead time of only eight weeks.

With increased productivity and reduced lead times comes enhanced customer service; both in terms of the responsiveness to customers in the design itself – for example, in designing in particular features – and in improving the speed of response to requests and invitations to tender. For example, Kaplinsky[17] quotes the case of a firm which was invited to tender for work in Mexico at 6.30 am on the day when their salesman was about to leave. By using CAD he was able to take all the necessary drawings and documents with him when the flight left only six hours later! Another firm increased its conversion of enquiries to firm orders by 10 per cent through the use of CAD.

CAD for the small company: two examples

Engineering

Even amongst small firms the benefits of CAD can be dramatic. An example which illustrates the new entrant experience is that of a small engineering firm in the Federal Republic of Germany, which specializes in making special-purpose machinery, usually on a one-off basis. These machines are complex, with a long build time. The firm's expertise and resources lie in design and assembly, so that much of their work is subcontracted out: hence there is a key role for design and drawing production. They employ about 14 staff and do not wish to expand the workforce by much more, preferring instead to use technology to improve productivity.

They are relative newcomers to CAD, having decided only about 18 months ago to move into this technology because of market pressures, and the desire to improve tendering by being able to present drawings as well as sketches at the early discussion stage. They spent a long time looking around the market, reading about and visiting possible installations. The company set no particular financial limit on their proposed investment but decided that large systems were not suited to their needs, primarily because they were overly powerful, sophisticated and capable for their needs, and were likely to be too complex to maintain and operate. They chose a PC-based system mainly on grounds of performance (within their required envelope) and expandability options.

Their main uses for the system are in making presentations at the tendering stage, as mentioned above, and in detailing and modification work. The actual design activity, in terms of translating concepts into engineering solutions, is still dependent on the designer's experience and skill, and the system has not made little impact on this – nor is it likely to.

Benefits to the firm have certainly justified the investment, and hold out great promise for future work. They estimate that without the system they would almost certainly have had to employ a second draughtsperson to handle the detailing on complex machines – so the investment can be considered justified on that basis alone. As indicated, the main benefit comes not in originating designs but in modifications and detailing, although the system has helped them to develop new ways of manu-facturing by providing accurate machining data rapidly enough to enable them to carry out complex progressive operations in real time. Other benefits have come in the improvement in the image which they present to customers and in the reduction in lead time; about 30–50 per cent of the lead time of 8–12 weeks for a machine is spent in negotiating and discussing changes with the customer, and so the CAD system has helped to reduce this considerably while improving the quality of the final design *and* customer satisfaction.

The overall costs for the system were about DM 40 000 and there is no maintenance contract. This is because the system uses standard and easily available components and the supplier, being local, is able to pro-vide a 'hotline' advisory service (although the system is very reliable). This cost needs to be seen in context; it represents only a small fraction of the typical contract value in the business of DM 200 000 – 700 000, and the system is fully utilized, unlike many higher-cost and more complex systems.

Mining equipment

A second example is that of a small firm making mining equipment and railway lines. Although an old-established firm, the business was bought out very recently and is now growing rapidly in a sector which has experienced considerable decline in recent years.

They invested in an Autocad system, and also found that it took very little time to learn the basics, and about six months to get up to speed. Its contribution has already been sufficiently impressive to justify the purchase of a second installation, and a further two workstations are planned to support a growing design operation. Significantly, part of their exploration and selection process was to look at what their competitors were doing: they found that, while some were not using CAD, others – notably the larger firms – were using large CAD systems with powerful capabilities but experiencing problems of utilization and optimal usage.

Their design task is fairly complex, involving a number of calculations, and CAD has reduced time on drawings by about 30 per cent – again mainly on the modifications rather than the origination stage. This is based on reductions from five hours to two hours for a typical drawing. As a result they estimate that without CAD they would not have been able to handle the current volumes of work.

In terms of training, the design staff are all basically draughtsmen by background and they have learned CAD on the job. Again, they have developed facility with programming within the system to enhance its usefulness to them.

The overall costs were about DM 90 000 for the two systems, including a large plotter. The current configuration is two Tandon 40 MB computers, two plotters, mouse digitizers and a large screen. Once again, no maintenance contract has been signed, but the standard nature of the equipment and the local dealer support is felt to be sufficient for a system which has already proved to be very reliable, and which they feel they understand. They estimate that the total operating costs are less than DM 1000 a year, largely made up of paper and pens, software updates, and so on.

Further benefits of CAD

However, although the above features are useful aids to the draughting process they do not really represent the full contributions which CAD can make. It also offers significant improvements in the *design* process, such as the following:

1   Improved quality of design, because the designer can simulate the final product more accurately and also because he has more time available to try different options (since redrawing is simply a matter of refreshing the display rather than a costly cycle of physical redrawing).

2   Further design quality improvements from the use of advanced simulation techniques to explore different aspects of the design; such as dynamic modelling ('What happens under different conditions?'), finite element analysis (to explore the pattern of stress and other variables across the whole product) and a variety of visual simulations.

3   Better integration of sub-assemblies into a complete design. For example, using such a system ensures that a modification to the engine will not conflict with the design of the bonnet of a car, because it becomes possible to display both systems working together.

4   The ability to relate to all other drawings under design at any moment in time. Once a three-dimensional electronic model of the product has been developed, any changes to any of the sub-designs can automatically be used to update the whole design. Thus the traditional problem of a large-scale development project, that of everyone working to different drawings because engineering changes are not communicated quickly, is eliminated.

5   Designs of much higher complexity can be attempted; for example, in VLSI chip production.

6   The intelligence of the system can be used to ensure conservation of expensive materials; for example, in the shoe industry or in sheet metalworking, where nesting programs are used to calculate the best pattern of pressing or cutting to minimize the cost of materials.

Perhaps the most significant benefit is the possibility of connecting the design process much more directly to the other parts of production. We have already seen that design provides information to the coordination area and to production; since these areas are also using electronic systems the obvious implication is to integrate them and share a common pool of information. This is the basis of CAD/CAM on the manufacturing side, and increasingly of CAPM systems on the coordination side.

CAD/CAM systems essentially take the output of the design process and convert the product dimensions to a series of instructions for the machines on which the product will be made. For example, this could be the cutting, drilling or other control instructions for a CNC machine tool program, or the control program to drive the automated looms making woven goods in a textile firm.

Since CAD systems hold all the information on what products will be made and what fixtures, tools, materials and so on will be needed, it is logical that they should be used to generate the basic information needed to order materials and schedule work for the factory – and such linkages between computer systems are increasingly common.

Finally, the picture is being painted of full computer-integrated manufacturing (CIM), in which a product is designed on a CAD system which then provides all the necessary information for both the physical production and the management of that process. In other words, the factory behaves as if it were a giant, complex but integrated machine.

Indeed, there is no reason why the process should stop at the boundary of the firm. Design information used between firms can also be brought into such an electronic network. This can either be within or between firms. In the former case, the use of design networks can link together design and engineering personnel on a worldwide basis, bringing extensive resources to bear and also allowing rapid updates of product data and specifications for all users. Ford recently announced a $77 million global network along these lines which will enable 20 000 engineers to exchange information. Based in Dearborn, Michigan, the WERS (Worldwide Engineering Release System) uses a standard format for storage and retrieval of a package of information relating to models and parts (of which there are currently some 700 000 in use). This is instantly updated to accommodate engineering changes made anywhere in the world – and thus represents a major resource to support design throughout the company.[18] Similar systems are in use in other companies and industries, for example telecommunications computers.

In the case of linkages between firms, the advantages come from allowing expertise in producing complex assemblies to be distributed amongst several companies. Thus instead of a manufacturer of a final product needing to be an expert on all the subsystems, these can be designed by specialist subcontractors, and the information relating to design can be built up on a co-operative basis. For example, in building a car the assembler produces some components and sub-assemblies in-house but buys many in from other specialist suppliers. The pattern here has been for these suppliers to install their own CAD systems. Therefore the potential exists for all suppliers to work on the same design model which is updated constantly just as the sub-designs are updated on in-company projects. The implications of this are for closer integration between suppliers and assemblers, especially in the area of R & D: this pattern is beginning to emerge in several industries, such as motor vehicles, aerospace, electronics and clothing.

7.6   BARRIERS TO EFFECTIVE IMPLEMENTATION

CAD clearly offers a powerful response to many of the demands for increased manufacturing agility and effectiveness that face firms in the 1990s. However, a number of problems still need to be solved before the kinds of benefit identified above can be exploited fully. We can also see, once again, the strong tendency to adopt CAD as a substitution innovation and to limit investment in more integrated and higher risk applications.

Two points emerge clearly from experience of CAD use. First, the information obtained from suppliers is often extremely optimistic with respect to potential gains and the extent of problems which may be encountered after installation. Second, to get the best out of the technology involves considerable organizational learning and adaptation. Most problems increase with the level of complexity in the CAD configuration and the degree of integration involved.

A 1984 survey of 1500 'turnkey' systems in the UK found four major sources of complaint:[19]

- the systems are too expensive

- the systems do not give the promised gains in productivity

- the systems take much longer than expected to master, especially to build up a database (suppliers' estimates of this time were considerably shorter than needed in practice)

- the systems provide very low levels of true CAD/CAM integration

While the cost problem may have subsequently diminshed, the other concerns are still frequently heard amongst CAD users. In another study, Francis[20] reported on four reasons for what often appeared to be inefficient or ineffective use of the technology:

- inappropriate systems for the task

- user fears, in some cases leading to continued use of manual systems in preference to CAD

- learning curve costs

- conflicts of interest – users prefer tailor-made solutions, customized to their particular problems, whereas suppliers want to sell standard packages

Martin et al.[21] found that maximal effectiveness of CAD was related to three factors:

- efficient application in context – obtaining the right system and fitting it in, not only to the design area but also with marketing, manufacturing and other functions

- systems integration
- levels of use of the system – which in turn depends on company-wide training, managerial as well as operator

Senker[22] reports on US research which identifies a number of reasons for the gap between investment expectations and actual performance:

- underestimation of the time needed to achieve effective functioning, often by a considerable margin
- overestimation of the utilization rate likely to be achieved
- underestimation of the need to make adaptive adjustments
- underestimation of the tasks involved in securing acceptance of changes by labour

Sawzin[23] draws an important distinction between short-term benefits which can be fairly easily achieved (such as reduction of errors, improved clarity, simplified revision of drawings, and elimination of drawing chores such as lettering or line work) and long-term benefits, which are where the main gains of CAD are to be found. Moving to the latter requires considerable organizational learning: estimates vary but most agree that it can take anything up to nine months just to learn to use CAD as well as a traditional drawing board. In their study of UK users, Martin et al cite one firm which had installed a suite of draughting and design programs that was justified on the basis of expected 5:1 productivity gains through reduced manpower and design times. In practice they reduced their drawing office staff by three draughtsmen, but although they were able to achieve considerable improvements in the short-term areas mentioned above they did not realize the longer-term benefits. They attribute this primarily to a lack of training.

It will be useful to consider some of these problem issues in a little more detail.

Technological problems

During the past five years significant advances have been made in the area of CAD hardware and software, but problems still exist at the point at which CAD systems integrate with other functions in the overall manufacturing chain. The case of post-processors to support CAD/CAM provides a good example of this. While it is theoretically possible to create a product design on a CAD screen and then pass control information direct to the machine tools involved in making it, in practice some form of conversion still remains to be carried out. The set of instructions to drive a CNC lathe, for example, will be different from those generated by the CAD software program and so an intermediate piece of software – a post-processor – will be needed to effect the translation. (This is not a trivial problem. Apart from product dimensional data a variety of other routines need to be incorporated in the final manufacturing control program, such as nesting calculations, optimal tool path calculations, fixturing data, and so on. These programs are often required to support rapid changes in product type/mix, as in a flexible manufacturing cell.)

Unfortunately there is, as yet, no universal language for either CAD systems or production equipment, so each item of equipment potentially requires its own post-processor. While CAD suppliers are increasingly offering post-processors for some of the more commonly used equipment as part of their sales package, there is still considerable effort required on the part of a user before full CAD/CAM integration can be achieved. Typical times quoted by software houses for the development of a post-processor for a specific application are around one week – an expensive additional investment. Even with post-processors supplied by CAD firms, options are usually incorporated for shop-floor editing of the NC programs, to allow for adjustment and modification during the manu-facturing process.

Networking and interfacing

With the increasing interconnection of CAD into other systems comes the need for some form of standards governing the communication of information within and between systems. A useful analogy can be drawn here with road traffic: unless there is an agreed and clearly defined set of rules and procedures governing the way in which traffic moves around the road system, the result will be chaos, and it will often be difficult or impossible to communicate between places.

By the same token, agreement needs to be reached about standards and protocols for moving information around electronic networks. Much of the discussion concerns the need for 'open systems' interconnection; that is, network standards which permit different items of equipment from different suppliers to be connected and to communicate. While this is a desirable goal, there are also obvious commercial advantages to be gained from supplying systems which do not communicate with every type of equipment, since this effectively excludes a customer from buying from competitors. The problem is compounded by the investments which CAD suppliers would have to make in converting their equipment to support new standards, and by the need to develop equipment – such as specialist network control chips – to facilitate the construction of suitable networks.

In the area of CAD a number of attempts have been made to agree standards, such as the Initial Graphics Exchange Standard (IGES) to permit communication between different CAD systems. In 1985 Boeing announced the development of a standard compatible with OSI guidelines called the Technical Office Protocol (TOP): this has been demonstrated working in conjunction with General Motor's Manufacturing Automation Protocol (MAP) to support the entire range of CAD/CAM activity from initial drawing through to final product.

In the UK, and internationally, developments in IGES to date have met with considerable success. Meanwhile, applications of MAP and TOP are limited to very large manufacturers, who have numerous pieces of equipment which need to be integrated. For the small company, the costs of MAP and TOP are still generally considered to outweigh the advantages.

Costs

The availability of low-cost CAD systems has undoubtedly helped to diffuse this technology, but it should be borne in mind that even at this level CAD requires additional investments in supporting facilities and in training. In addition, there can be substantial learning costs, as Arnold points out: '. . . it is difficult to escape learning costs in adopting CAD which is an inherently productivity-enhancing and therefore labour-displacing technology applied within an already existing pool of labour. Under these circumstances training for CAD involves retraining draughtsmen.'[24] In their survey of 34 UK users, Senker and Arnold found that it took an average of two years for firms to achieve best-practice productivity levels of the kind suggested by the suppliers at the time the system was installed. However, the costs of this learning process – both in terms of the direct training costs and the cost of the 'lost' productivity – were rarely included in cost justifications at the investment proposal stage.[25]

Although low-cost CAD is available on microcomputers, the full power implicit in CAD/CAM and CIM-type implementations requires a much more substantial investment. In turn this poses problems for some firms because of the difficulty of justifying investment based on simple payback criteria, such as improvements in draughting productivity or savings in draughting labour. In many cases firms indicate that their justification has been something of 'an act of faith'. As Senker and Arnold[26] point out, such methods of justification continue to be used largely because they represent a formula which is conventionally understood in investment appraisal. The real benefits from CAD come not from labour-savings – and only to a limited extent from draughting productivity improvements – but from changes in the way the design process operates and consequent savings in lead times, increases in responsiveness and the overall contribution that these factors make to competitiveness.

In the context of US users, a similar point is made by Beatty and Gordon, who criticize what they see as an excessive focus on direct labour productivity rather than effectiveness:

> ✗ CAD/CAM can and should dramatically alter the way a company
> carries out its tasks. If used merely to imitate existing processes, it
> will remain both expensive and ineffective . . . to use the tech-
> nology successfully, companies have to rethink the way they
> operate and question previous rules of thumb. Among the most
> important of these is the focus on direct labour and the rigid
> adherence to manufacturing ratios[27]

Such attitudes lead, in their experience, to 'dysfunctional' behaviour: for
example, with managers creating a formal justification based on
traditional measures just to "fool the beancounters" rather than a realistic
case for a strategic move forward in technology.

Skills and training

In their major 1982 study of the training implications of CAD, Senker
and Arnold[28] suggested that there was a substantial need for training at
several levels in order to exploit the full benefits of CAD. In their sample
firms reported that it took up to two years on average before designers
and draughtsmen were sufficiently proficient with their systems to
achieve high productivity. Follow-up work has monitored the progress of
15 of the firms studied during the mid-1980s, when so much change and
expansion in CAD technology took place.[29] Significantly, the main
findings of this research still emphasize the pressing need for training
and skills development at all levels:

> Overall it is apparent that commitment to CAD and CAE training
> in the industry is far from adequate to ensure the effective use of
> technology. Managers, especially senior managers, are not yet
> sufficiently committed to training. The most grievous deficiency is
> in the training of senior managers.

Other evidence confirms that the training requirement is not confined to
operating skills alone; in a study of West German systems one of the major
reasons for failure or poor performance was that the people in charge had an
inappropriate background or no experience in design work, and did not
understand the importance in the design process of key elements – for
example, the concept of function in design as opposed to artistic elegance.[30]

Another important finding was the key role played by 'project cham-
pion' figures in the CAD innovation process. This role – which the team
termed 'CAD-mother' – involves not only pioneering and championing
the investment decision and the early implementation but also the later
stages of CAD support, including training, design office reorganization
and long-term system support and development.[31]

The extent to which training is required by users varies with the type
of system installed. A designer/draughtsperson can achieve basic

draughting competence on entry-level low-cost systems in only a short time – a matter of days to weeks in most cases. This is partly a function of the relative simplicity of the tasks involved and partly a result of user friendly techniques (such as menus and pull-down windows) to assist self-teaching work. For this reason little if any training is usually offered by suppliers of such systems, although tutorial and 'hotline' support is usually provided.

As complexity in systems or applications increases, so does the need for higher and more specialized skills. Thus, the extension of a low-cost CAD system to more advanced applications generally requires more time; typically about six months to master applications programming within a high-level language such as LISP. Larger and more powerful systems are usually supplied with some training as part of the elementary package, but this normally covers basic system operation only. Once again, the move to more specialized work towards exploiting the full capability of the system requires further training inputs after initial facility has been achieved.

Inter-firm linkages

With the increasing complexity of many products comes the need for extensive design co-operation, particularly in areas where specialist expertise and experience may be needed. In theory, the networking of CAD between firms should not pose a problem, provided that there are suitable standards governing the exchange of data. This pattern is already beginning to emerge in industries such as motor-vehicle building, where the main assembler acts as a focal point and final consumer for designs for a variety of components and subsystems produced by a wide range of firms.

This process was traditionally cumbersome, costly and inaccurate, with a limited exchange of information. For example, in the development of a motor car it was typical to find upwards of 20 000 different drawings for various parts and sub-assemblies. Every change which the final assembler or the component supplier wished to make involved a lengthy exchange of information and a laborious updating of drawings. A number of problems are associated with this approach. First, the procedure is difficult and so the tendency is towards sub-optimal designs. Second, the fact that – at any one time – many different design changes are being carried out means that it is very difficult for each participant to be sure that they are working on the same drawing; and this can lead to problems in cases where the systems or components interact.

With a CAD-linked system there are immediate benefits for the quality of the design, the currency of the drawing and the scope of the possible experimentation. In the case of such inter-company developments the concept of 'paperless' engineering is now possible. A component (or indeed, the entire car) can be developed jointly by the vehicle manufacturer and the component supplier without the need for conventional drawings. For such levels of co-operation, however, the trust between the two parties must be based upon long-term, open relationships, which are perceived by both as a path towards mutual benefit. There seems little doubt that this will be the pattern for the future but, as Lamming has indicated elsewhere, successful exploitation of such links will depend critically on changing the basis on which firms relate to each other.[32] Traditional confrontational approaches need to give way to long-term co-operative ventures.

Although progress is being made in establishing communication standards for CAD networks, there is still no universally recognized or technically proven system available. This has major implications for projects which require sharing of design information between firms, since the CAD systems in use may be incompatible. Some indication of the difficulties at a technical level in getting different CAD systems to talk to each other within a supplier network can be seen in the results of a survey carried out by a major West German car manufacturer.[33] Of 49 key suppliers examined:

- 24 were directly compatible via IGES

- 14 were incompatible via IGES but compatible via an alternative standard developed by the vehicle manufacturer's trade association (Verein der deutscher Automobilindustrie)

- 7 were capable of exchange of data via either standard

- 16 were complete CAD 'islands', unable to exchange or communicate in any form with the assembler

Organizational integration

In most firms the product development process involves a number of functions, ranging from marketing through design and development to manufacturing, quality assurance and finally back to marketing. Differences in the tasks which each of these functions performs, in the training and experience of those working there, and in the timescales and operating pressures under which they work, all mean that each of these areas becomes characterized by a different working culture. Functional divisions of this kind are often exaggerated by location; for example, R & D and design activities are grouped away from the mainstream production and sales operations – in some cases on a completely different site. This basic division by function is indicated in table 7.1.

Separation of this kind can lead to a number of problems in the overall development process. Distancing the design function from the

Design/drawing office	Manufacturing	Marketing
Concerned with the aspects of the product itself – technical excellence, and so on	Concerned with how to meet production targets for making the product	Concerned with how to respond correctly and successfully to the market
Wish to be able to revise plans repeatedly to exploit technical advances	Wish to fix design and limit subsequent changes to essential modifications only	Wish to alter each product to precise needs of each customer and exploit technology
Like to design each part of each new product anew	Would like to standardize component parts	Wish to customize products, regardless of components
Concerned with specifications	Concerned with timing	Concerned with sales figures

Table 7.1   Differences in attitudes to the design process

marketplace can lead to inappropriate designs which do not meet the real customer needs, or which are 'over-engineered', embodying a technically sophisticated and elegant solution which exceeds the actual requirement (and may be too expensive as a consequence). This kind of phenomenon is often found in industries which have a tradition of defence contracting, where work has been carried out on a cost-plus basis involving projects which have emphasized technical design features rather than commercial or manufacturability criteria.

Similarly, the absence of a close link with manufacturing means that much of the information about the basic 'make-ability' of a new design either does not get back to the design area at all, or does so at too late a stage to make a difference or to allow the design to be changed. There are many cases in which manufacturing has wrestled with the problem of making or assembling a product which requires complex manipulation, but where minor design change – for example, relocation of a screw hole – would considerably simplify the process. In many cases such an approach has led to major reductions in the number of operations necessary; simplifying the process and often, as an extension, making it more susceptible to automation and further improvements in control, quality and throughput.

It is important to place this in perspective. Estimates suggest that up to 70 per cent of the cost of a product is determined at the design stage, and thus it would seem important to concentrate attention on hammering out any likely problems in manufacturing at this stage. Yet most firms spend less than 5 per cent of their product budget on design and, instead, push for manufacture as soon as possible. While this may reduce the time-to-market, it can also have a major effect upon costs. A variety of studies have consistently shown that attention to product simplification and design for manufacture can lead to substantial savings at later stages.[34] For example, in a study of the IBM Lexington plant for computer printer manufacture, simplification of design led to a reduction of 3:1 in part numbers, and this in turn meant a cut of 9:1 in adjustments.

Once again, there appear to be important lessons in Japanese practice for the organization and management of design. Lorenz[35] and Scibberras and Payne, amongst others, point to a number of factors which are of significance in the Japanese model. Central to their success was the concept of what could be termed *integrated design*. Very close links were forged between product designers, marketing and manufacturing representatives, such that the final product was the result of a continuous dialogue throughout the design and development process In contrast, the traditional Western approach is essentially sequential in nature.

The need for a more integrated approach has been recognized for some time. For example, Sir Kenneth Corfield, in the summary of his 1979 report on product design in the UK, commented:

> Design should be carried out in a 'multi-disciplinary' way: all relevant functions of the business should be brought to bear on all stages of product development, especially marketing, production and finance. Too often the design of a product is developed in isolation from all the other key functions. It has then to be adapted at a relatively late stage of its development to meet market requirements that were inadequately identified at the outset, or to meet the production process needs, or to keep within the cost budget.[36]

7.7   SIMULTANEOUS ENGINEERING

The idea of interplay between designers, makers, sellers and users is not new. Indeed, it provides the basis for the product improvement process which operates today in which marketing – in touch with end users – report back to the design function information about problems, suggested improvements, and so on. Von Hippel and others have drawn attention to this as an important source of innovation.[37] Similarly, the experience of making the product on the shop floor eventually finds its way back into the drawing office and leads to suggestions for improvements.

But this is often a slow and reactive process which, it can be argued, is not very efficient as a mechanism for improving design performance in an era of shortening product life cycles, and increased competition on the basis of non-price factors. By contrast, an integrated approach, such as that suggested above, is essentially a proactive one in which the design is refined and developed on the basis of 'real-time' interaction so that it is constantly evolving and improving. The benefits of such integration include:

- improved 'makeability'

- reduced costs of production (labour and equipment) through simplification of the process

- a reduced floorspace requirement for production through simplification

- improved use of materials (and thus reduced material costs)

- reduced set-up times (and thus small-batch capability, flexibility, and so on) through definition of product families, commonization of several similar parts into one universal part

- reduced costs of tooling, fixtures, and so on

- improved product design and manufactured quality

- reduced production lead time

- reduced design and development lead time

- cumulative benefit of improved communication between departments

- reduced time-to-market for new products

Considerable interest is now being shown in the idea of 'simultaneous engineering' as a means to ensure a more rapid time-to-market for new product developments. Estimates suggest that being first into the market means that a firm obtains a 50 per cent market share for that product or service, and other benefits include reduced costs (of work hours and inventories) and improved customer relationships because of better, more rapid service.[38] Examples of firms which have made dramatic cuts in their time-to-market include Honda (which cut its five-year new car development cycle by 50 per cent), ATT (which also reduced by 50 per cent the time taken to introduce its new cordless telephone) and Hewlett–Packard (which reduced the time to develop its printers from 4.5 years to 22 months). The Xerox Corporation has been making dramatic reductions in development times, cutting them from six to around three years on a range of office products, as a consequence of the competitive pressure from Japanese manufacturers.[39] 'Cycle compression' is another phrase which is beginning to emerge, especially in the context of long development cycle products such as military equipment and aircraft.

Simultaneous engineering aims at four key targets:

- to reduce the time lag between different stages in the traditional sequence of product development, marketing and sales

- to minimize development and final product costs

- to increase product quality

- to increase market share through early entry

It is based on a simple principle which involves moving away from what might be termed the traditional 'relay race' approach, with each player in the process of passing the baton (the new product) on to the next, but seldom discussing the overall progress until after the race. Instead, in simultaneous engineering, teams representing all disciplines work through in parallel. The main advantage in such an approach is the early identification and resolution of conflicts – it prevents 'passing the buck' and demands a co-operative solution. Better information flow also brings in new inputs to design at a stage where they can be used to make improvements – as distinct from the traditional practice of apportioning blame to different areas because they failed to pass on information which could have helped to avoid costly design faults.

Lamming and Bessant[40] provide an interesting illustration of this alternative approach in the motor industry. UK companies involved in joint ventures with Japanese partners have sometimes complained about the lack of detail on drawings sent from Japan, assuming that correct product quality can only be achieved through precise (often pedantic) specification at the design and draughting stages prior to "release" to production. The Japanese method of early (and therefore quick) release from design to production, followed by continuous development of the product by production departments, has yielded three major areas of benefit, however, which have gradually been realized by the UK partners. First, modification of the product in production ensures continuous improvement in process efficiency (and thus reduction in costs, improvements in quality, and so on). Second, design improvements suggested by production have a natural path for communication, and may be incorporated with a minimum of difficulty. Third, early release of the product enables the designers to concentrate on the replacement model, thus reducing development periods by eliminating 'afterthought'. The apparent 'sloppiness' of some Japanese engineering information (drawings, for example) is in fact the sign of a more open design authority, which is shared for mutual benefit (between departments) rather than jealously guarded for individual 'professional' security. The technique is sometimes referred to by the Japanese as using the factory as the laboratory of the designer.

Such a teamwork and co-operative approach needs to be set against the traditional structure of design, which is often seen as a distant function, pursuing very different goals to those of the rest of the organization. Successful teams stress mutual learning; indeed, some companies prefer to make continuing use of such teams since the investment in group development is so valuable. They need a clear goal towards which they can all work – and the leadership needs full authority to challenge what has traditionally been done. For example, in a report on the development of Nissan's Maxima car, the project engineer was able to preserve his design team authority over the heads of board-level management. Such an approach avoids costly last-minute changes to placate senior management – a major difference from US counterparts, where senior management often has extensive influence on the process. However, the benefits of the Nissan approach are beginning to emerge; the Maxima was developed within 30 months and also won the industry's top quality award soon after its introduction – and in 1989 it was ranked as the most trouble-free car in the US by users.[41]

In many ways, this is the other side of CAD/CAM technology. As Winch points out, computer networks enable integrated design to take place, permitting a two-way flow of information, instant updating, and so on. But on their own they cannot guarantee closer working together towards the same objectives.[42] A degree of integration is essential to make CAD work at all, but much more can be gained by taking a fully integrated organizational approach. Once again, there is a need for significant organisational adaptation in order to obtain the full benefits offered by new technology.

Notes

1   See, for example, A. de Meyer and K. Ferdows, *Quality Up, Technology Down*, Working Paper No. 88/65, INSEAD, Fontainebleu, France, 1988.
2   R. Hayes, S. Wheelwright and K. Clark, *Dynamic Manufacturing*, (Free Press, New York, 1988).
3   K. Corfield, *Product Design*, (National Economic Development Office, London, 1979).
4   *The Engineer*, 19 February 1965.
5   An excellent description of this period of technological innovation can be found in E. Arnold and P. Senker, 'European prospects in the CAD industry', *CAE Journal*, October 1985, pp.150–6.
6   E. Arnold, *CAD in Europe*, Report No. 6, Science Policy Research Unit/Sussex European Research Centre, Sussex University, 1984.
7   *Industrial Computing*, October 1987.
8   'CAD's fourth generation', *The Engineer*, 27 October 1988, p.32.
9   *Engineering Computers*/Benchmark Research 1988 survey.
10 J. Northcott, W. Knetsch and B. de Lestapis, *Microelectronics in Industry*, (Policy Studies Institute, London, 1987).
11 US Bureau of the Census data, 1988.
12 *Engineering Computers*/Benchmark Research survey – see note 9.

13 J. Bessant, *CAD in the Federal Republic of Germany*, Centre for Business Research, Brighton Business School, 1986 (mimeo).
14 R. Kaplinsky, *Computer-aided Design* (Frances Pinter, London, 1982).
15 E. Arnold and P. Senker, *Designing the Future: the Skills Implications of Interactive CAD*, Occasional Paper 9, Engineering Industry Training Board, Watford, 1982.
16 Kaplinsky, *Computer-aided Design*.
17 Kaplinsky, *Computer-aided Design*.
18 *Automation*, February 1989, p.5.
19 *New Technology*, March 1984.
20 A. Francis, 'The human side of CAD/CAM', *CAD/CAM International*, December 1984.
21 R. Martin et al., 'Some human factors in effective CAE', *CAE Journal*, February 1985.
22 P. Senker, 'Implications of CAD/CAM for management', *Omega*, 12 (3) 1985.
23 S. Sawzin, 'The design of a computer graphics training programme', *IEE Computer Graphics and Applications*, 3, 1983.
24 E. Arnold and J. Bessant, 'Oiling the wheels of technical change', in *Training for Tomorrow*, ed. J. Rijnsdorp (Elsevier, Amsterdam, 1983).
25 Arnold and Senker, 'Designing the future'.
26 Arnold and Senker, 'Designing the future'.
27 G. Beatty and J. Gordon, 'Barriers to the implementation of CAD/CAM systems', *Sloan Management Review*, 25 (1988), pp.25–33.
28 Arnold and Senker, 'Designing the future'.
29 P. Simmonds and P. Senker, *CAE in the 1980s: a Report on a Longitudinal Study in British Engineering Companies*, Science Policy Research Unit, Sussex University, 1989 (mimeo).
30 M. Rader, B. Wingert and V. Riehm (eds), *Social Science Research on CAD/CAM*, (Physica-Verlag, Heidelberg, 1988).
31 Rader, et al., *Social Science Research a CAD/CAM*.
32 R. Lamming, ' For better or worse? Buyer–supplier relationships in the UK motor components industry', in *Managing Advanced Manufacturing Technology*, ed. C. Voss (IFS Publications, Kempston, 1986).
33 Private communication.
34 'Less is more', *Manufacturing Engineer*, November 1989, p.20.
35 C. Lorenz, 'Seizing the initiative in a struggle for survival', *Financial Times*, 17 June 1987.
36 Corfield, *Product Design*.
37 See, for example, B. Shaw, 'Strategies for user/producer interactions' and G. Foxall, 'Strategic implications of user-initiated innovation', *Innovation, Adaptation and Growth*, ed. R. Rothwell and J. Bessant, (Elsevier, Amsterdam, 1987); and E. von Hippel, 'The dominant role of users in the scientific instrument innovation process', *Research Policy*, 5 (3) (1976), pp.212–39.
38 'United front is faster', *Management Today*, November 1989.
39 C Lorenz, 'Seizing the initiative' in a struggle for survival, *Financial Times*, 17 June 1987.
40 R. Lamming and J. Bessant, 'Design for efficient manufacture', in *International Handbook of Operations and Production Management*, ed. R. Wild, (Cassell, London, 1988).
41 'Bosses who stay out of the driving seat', *The Guardian*, 12 February 1990.
42 G. Winch, 'The implementation of CAD/CAM:concepts and propositions', *Warwick Papers in Management*, Number 24, September 1988.

8 Leaner and Fitter: the Just-in-time Approach

8.1 INTRODUCTION

In the preceding chapters we focused on applications of IT in its more integrated forms, and explored some of the ways in which they help to deal with the challenges of competing strategically in the 1990s. The cases presented in 'before and after' format in table 8.1 illustrate suc-cessful performance along these dimensions.

Factor	Before	After
Company A		
Throughput time	25 days	2 days
Inventory turns/year	5	30
Delayed deliveries	40 per cent	2 per cent
Inventory cost	£10 million	£2 million
Hours/unit (all staff)	330	200
Rework	6 per cent	1 per cent
Company B		
Lead time	10 days	1 day
Work in progress	15 000 units	1000 units
Rejects	3 per cent	1 per cent
Inventory turns/year	25	200
Overall costs	—	Cut by 10 per cent

Sources: A Ingersoll Engineers, 1986[1]
 B author's research

Table 8.1 **Performance improvement through organizational change**

189

However, there is an important difference between these examples and those presented in earlier chapters. All of these have been achieved *without* investment in advanced applications of IT but simply through making changes in the way production was organized and managed. Central to these success stories has been the adoption of alternative management practices, which originally evolved in Japan but which have now received much more widespread application. In particular, emphasis has been placed on two approaches which we shall explore in this chapter and the next – just-in-time manufacturing and total quality management.

The effectiveness of just-in-time and total quality management in dealing with the challenge of competitiveness should not come as a complete surprise to us. After all, we saw in earlier chapters the repeated emphasis on the need for organizational change to support the effective introduction of AMT. And we also saw that in many cases users attribute much of the benefit to these changes rather than to the actual technology which they introduced. The main thrust of these changes has been to challenge the traditional models of manufacturing organization and management, which evolved in an era of stable markets and mass production.

It is important to see them not as alternatives but rather as comple-mentary to advanced manufacturing technology. As has already been said, technology is more than just an arrangement of machines and computer systems. It also embraces the people, skills and work organization used to achieve a particular manufacturing goal – and it is in this area that just-in-time and total quality management represent effective innovations. But because they are about disembodied change rather than physical equipment it is often difficult to understand them fully or to effect change using them.

This chapter will focus on just-in-time manufacturing (JIT). JIT was born in Japan out of concern to reduce waste, and while it has become particularly effective as a weapon against excess inventory it is much more than that. In true Eastern style it is a philosophy – and often misunderstood and easy to dismiss in this form. But evidence is mounting that it is an extremely powerful integrated system innovation, much more than a technique for improving materials management.

8.2   NEED PULL: ATTACKING THE SEVEN DEADLY WASTES

After the Second World War Japan faced a series of challenges; shortages of resources of all kinds, a severely damaged and underdeveloped industrial base, restricted access to world markets and a reputation for poor quality. Yet within 50 years Japan has become the dominant industrial power in the world, able to attack, capture and successfully defend markets in fields as diverse as shipbuilding and microelectronics, motor vehicles and biotechnology. Its average annual growth rate of manufacturing productivity between 1960 and 1980 was 9.3 per cent, compared to only 2.7 per cent in the US. The trend in the automobile industry is highlighted in figure 8.1.

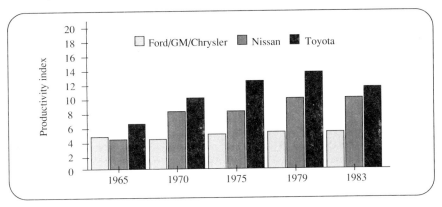

Figure 8.1   Productivity growth in the Japanese automobile industry. Productivity index = vehicles per year adjusted for vertical integration, capacity utilization and length of working year.
Source: **Cusumano, 1985**

Japan has also continued to adapt and cope with challenges arising from a lack of natural and energy resources, overcrowding of a large population on to four islands, external shocks such as the 1973–4 oil-price crisis and problems resulting from its own success which has led to a highly valued yen.

Part of the process whereby this has been achieved reflects on the alternative approaches which have evolved to organizing and managing production. In a country with scarce resources it is not entirely surprising that there was a considerable preoccupation with elimination of waste and with careful conservation – a practical concern which was underpinned by a deeper cultural set of values concerning thrift and economy. Schonberger summarizes the approach by attention to 'the three Ms – 'muri' (unreasonableness), 'mura' (unevenness) and 'muda' (waste)'.[2]

'Waste' is a broad term, one definition of which is:

> . . . anything that adds cost, not value . . . anything other than the minimum amount of resources absolutely essential to meet customer requirements.

As Taichi Ohno, one of the 'founding fathers' of JIT, put it:

> . . . there is nothing more wasteful than producing something you do not need immediately and then storing it in a warehouse. Both

people and machines are wasted and the warehouse puts your money to sleep.[3]

It is useful to reflect on the various different kinds of waste which can be found in most factories. Shigeo Shingo, another of the architects of the JIT approach, identifies seven areas in which there is waste:[4]

1 *Overproduction*, where the plant or stages within it make more than is actually necessary to fulfil an order or supply the next stage in production. The causes of this may be a desire to keep machinery utilized, or to try to offer better customer service by holding a high level of finished goods in stock. However, the costs associated with such policies can be serious: they include not only costs of working capital tied up in inventory and associated interest charges but also the cost of extra storage, handling, paperwork and people to monitor and control it.

2 *Waiting time*, where parts or products are waiting for the next operation, machines and operators are waiting for the next batch of work to arrive, or where machines and operators wait for specialist support (maintenance, quality control, and so on). Such delays, which are typical of batch manufacturing operations, represent not only wasted time but also imply inefficient flow, with high levels of inventory locked up wastefully within the system.

3 *Transport*, where parts and products are handled and moved around the factory more than necessary. For example, in many factories it is common for components to travel many kilometres in the course of their assembly into finished products because of the layout of operations. This kind of problem can occur in factories which are only a couple of hundred metres door to door. Another associated problem is putting things into temporary storage and then having to retrieve them – which adds not only to double handling and transportation waste but also to waiting time in the plant.

4 *Processing waste*, in which the actual process used may be wasteful or inefficient and could be improved upon or replaced. Here the question of maintenance and design for manufacture can make a major contribution to reducing processing steps or finishing operations. For example, a complex assembly operation can be simplified through better design of fixtures and of the product for manufacture, thus reducing the overall processing time, complexity and error rate. Similarly, well maintained tooling can reduce the time spent on, or even the need for, finishing operations.

5 *Inventory*, which, as we have seen, represents a major cost item. High inventories arise from overproduction but also from other wasteful policies and practices, such as buying high volumes of

raw materials at a discount and keeping them in stock, or holding on to obsolete materials or retaining too wide a product range for too long. The problem of inventory can be put into perspective if we consider that around 40–50 per cent of manufacturing costs (more in some industries such as leather or specialist metals) are represented by materials.

6   *Quality*, where the presence of errors and defects leads to physical waste in the form of scrap, time waste in the form of dealing with the problem or reworking, inventory waste because of the need to hold more stock to cover for the defective elements, and so on.

7   *Motion*, where the problem is that movement does not necessarily mean productive activity. Keeping a machine running may lead to high utilization figures, but if it is producing more than is actually needed the advantages of high utilization may be outweighed by the costs of extra inventory. Another waste of motion is in searching for tools and other items needed to complete an operation.

The extent to which competitiveness can be improved by eliminating waste can be seen by considering the high costs currently associated with inefficient use of inputs to production. For example, estimates suggest that the amount of money tied up in inventory (raw materials, finished goods, work-in-progress, and so on) in the UK alone represents between £23 billion and £41 billion.[5] Similarly, estimates suggests that some 30–40 per cent of energy could be saved through more attention to conservation. Or again, paperwork systems in many firms require forests of trees to keep them supplied, quite apart from the costs associated with the people involved in operating such systems. Here again, rethinking and simplifying procedures could result in significant savings.

JIT was born out of a desire to address this problem by aiming at producing things *just in time* for them to be used; that is, with the absolute minimum of waste. This is clearly an ideal, but it is a target towards which continuous improvements can be directed. The extent to which Japanese firms have been successful can be seen if we compare the stockholding costs of Japanese and US automobile producers. Estimates suggest that it costs US manufacturers around $8.5 billion to carry stock whereas the total figure for Japanese industry is only $800 million.[6]

In this form JIT is a simple philosophy, equally applicable to activities within the factory, dealings between factories along the supply and distribution chain and even in many service sector activities. Its main emphasis is on identifying where waste exists, in whatever form, and focusing on a variety of problem-solving strategies to deal with it.

The primary emphasis was originally in the area of inventory management. In Japan, with little or no natural resources, conservation was critical: but the problem was – and is – just as acute in Western countries. However, the Western approach has traditionally been to use inventory as a way of dealing with uncertainty rather than as something to be reduced or eliminated. For example, on raw materials the lead times and reliability of suppliers are often less than ideal and so the habit forms of holding safety stocks, 'just in case' of problems. The size of such safety or buffer stocks varies, but can often be 50 per cent or more of the total inventory. Uncertainty on the factory floor, associated with unexpected problems of machine breakdown or delay, is often dealt with by starting new batches and rerouting or rescheduling around the problem – and again inventory accumulation is used as a cover. And uncertainty in the marketplace, about what customers want and when they want it, coupled with a desire to provide good service to customers is again dealt with by holding high inventories of finished goods.

Inventory is being used here as something to cover up or cope with a wide range of problems in the factory and marketplace. If some of those problems could be solved, uncertainty would be reduced and the need to hold so much inventory would fall. But most Western manufacturing operates on the basis of crisis management: dealing with problems as they arise rather than systematically working down to their causes and eliminating them. And even when a problem is attacked, it is often the wrong problem: for example, our preoccupation with direct labour-saving whereas the main cost item for most manufacturers is the cost of material.

Clearly, we need a way of focusing on these waste-related problems in such a manner that action to solve them is forced upon the firm. This principle is at the heart of JIT. A widely used analogy, that of inventory as water in a lake or river, helps to explain this (see figure 8.2). For as long as the level is high, the rocks and other obstacles remain hidden, but as soon as the level is dropped, the problems emerge and can – and must – be attacked directly. JIT begins by identifying problems and then forcing firms to tackle them. The main tactic used to 'reveal' such problems is inventory reduction.

The way in which Japanese manufacturers attacked the inventory problem was essentially to challenge some of the conventional wisdom of batch manufacturing. In particular, they focused upon the idea of producing in response to need rather than as a consequence of plans and forecasts. Instead of pushing inventory into the system in order to make products they turned the process round and used the pull from the marketplace or the next operation as a way of making the system more directly responsive, and eliminating unnecessary waste due to over-production, and so on.

In 'conventional' production planning the idea is to convert customer orders to a plan and then procure the necessary materials in order to meet

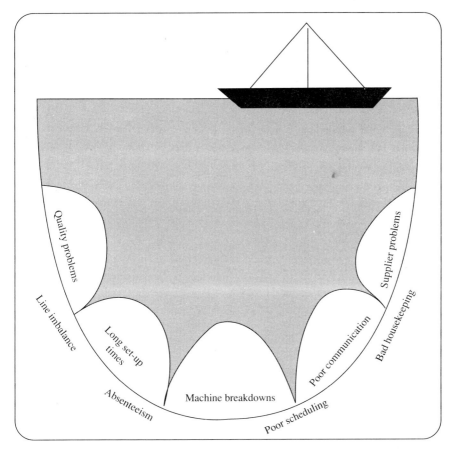

Figure 8.2 The river of inventory

that plan. Stocks of raw materials are pushed in at one end and products come out of the other. When problems crop up, more material is pushed in. Ways of controlling and optimizing such push systems, although available in various often complex CAPM systems, are the basis of traditional approaches. By contrast, JIT works on the principle of pulling through from the next user or customer: nothing happens until just in time for it to be used. And in doing so it represents a major risk: unless the entire production system (and its extension into the supply chain) is highly responsive, then the only effect of trying to operate JIT will be that the factory quickly stops producing as it runs out of material or a machine breaks down. Working with low inventories *demands* that problems are solved, not hidden.

Of course, some businesses are geared up to make to order – custom clothing, for example – but others essentially make for stock in large batches. Equally, some are used to working with low inventories; for example, baking runs on low flour inventories and frequent deliveries just in time for the next batch of bread to be made. But in all of these there is still considerable room for improvement in overall responsiveness.

The nature of this process of challenging 'conventional' wisdom can be seen in the approach taken to the classical economic batch quantity (EBQ) theory. As we have seen in earlier chapters, batch manufacture poses problems because of the 'productivity dilemma'; the more variety offered, the less time the plant is actually producing because of the problems of interruptions, breakdowns, resetting, and so on. Thus, all other things being equal, manufacturers wish to produce the longest possible run of a single product before they have to stop and change over.

EBQ theory developed as a mathematical response to the question of how large a batch firms should make before resetting so as to still operate economically. The theory involves trading off the costs of holding inventory (the 'carrying cost', which increases with increasing volume of production), the cost of producing (which decreases with volume) and the costs of resetting (made up of the time lost, the costs of paperwork associated with processing a new batch of orders, and so on). It is usually expressed graphically, as in figure 8.3, where it can be seen that there is a batch size at which the best balance between these is achieved. This EBQ is used to schedule the batch sizes in manufacturing. It does not mean that all the products made in that batch will be sold, but simply provides an indication of the most efficient use of production resources.

However, this is production-led – not market-led – and thus leads to overproduction.b The problem is further compounded by the natural interruptions and breakdowns in production, which are unpredictable. As

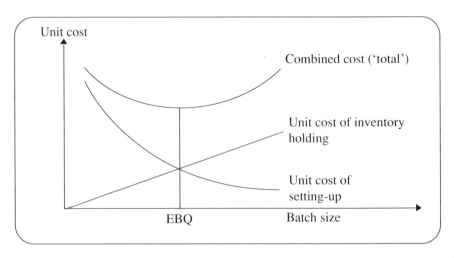

Figure 8.3   Economic batch quantity

we have seen, these are often dealt with by pushing more inventory into the system. The principle becomes 'if one batch can't be completed, then make some of something else and hang on to the original batch until it can be finished' adding to WIP and also to overhead costs in keeping track of it. The effect of a push system such as this is that it is possible to satisfy the customer, but at a high cost of inventory carrying.

JIT begins to challenge the underlying assumptions of EBQ theory. In particular, the idea of a fixed cost of setting up only holds for as long as there is no way of reducing set-up times. If these can be brought down, then the whole pattern of economic batch size changes – and it becomes economical to make in smaller batches.

It is important to note its origins in the car industry. Post-war Japanese producers had no choice but to try and evolve a more efficient way of producing in small batches, because the size of the domestic market was too small to permit efficient use of Fordist principles. Mass production volumes (which were estimated to require a volume of around 250 000 cars/year) did not emerge until 1959.[7]

JIT takes this to its logical conclusion and aims for a set-up time low enough to permit economical manufacture of a batch size of one. This is clearly an ideal, but any progress along the road to achieving it can make a huge difference to inventory management and, indeed, to dealing with many of the other wastes. Nor does making in batches of one preclude high-volume manufacture – the production series is simply of the same product. Indeed, one of the goals of JIT is to make batches repetitively so as to approximate the smooth flow of the continuous process industries. But adopting this approach to higher-variety manufacture opens up an opportunity to resolve the 'productivity dilemma', since there is no difference between batches of one of the same thing or of totally different things.

What this highlights is that JIT is fundamentally about developing manufacturing *flexibility*, the ability to switch production quickly with minimal delays. The essence of the need pull which brought JIT innovation into being is thus how to operate with a batch size of one, pulled through from the next customer or process stage. This challenge is, of course, easy to state, but to achieve it in practice requires the solution of a whole range of problems, some general and some highly specific to particular firms – which returns us to the notion of JIT as an approach based on 'enforced problem-solving'.

For example, to meet the JIT challenge of producing in batch sizes of one, some way is needed to:

- reduce set-up times

- guarantee that materials will be there without holding excess inventory

- guarantee machine availability and reliability

- ensure easy and rapid availability of tools and fixtures

- guarantee incoming quality with zero defects

- ensure smooth flow through the plant

- reduce inventory without the risk of running out of stock

- make the whole plant responsive and agile

The analogy of water introduced earlier is a powerful illustration of this. As we drain the level so more rocks become exposed, just as in the factory the process of reducing inventory levels reveals problems which must be confronted, not buried again. For example, if the problem is machine breakdown, then some way has to be found – replacement, special maintenance or whatever – to improve its performance. If it is set-up time, then some way has to be found of reducing it. Two principles are important in this process of continuous problem identification and solution. The first is that once one problem is solved, there is always another one to take its place. And the second is that there is no such thing as a 'best' solution – only an opportunity to find an even better one.

8.3   TECHNOLOGY PUSH – HOW TO CLEAR THE ROCKS

Clearly, the solution to these problems did not appear overnight. The process took around 30 years, and still continues. It is generally acknowledged to have begun in the 1960s in the Japanese car industry, with the Toyota company taking an important lead. But while it is possible to identify several common elements, JIT still remains a company-specific approach; indeed, it is not even called JIT in Japan, but instead is known as the Toyota Production System, or the Maxell Minimum Stock System.[8]

What is JIT? Schonberger defines its purpose as being:

> . . . to produce and deliver finished goods just in time to be sold, sub-assemblies just in time to be assembled into finished goods, fabricated parts just in time to go into sub-assemblies and purchased materials just in time to be transformed into fabricated parts.[9]

Consequently, it is a *total system* approach rather than one targeted at a particular area of manufacturing – in our terms, an integrated rather than a substitution innovation. It offers an umbrella approach to dealing with problems, especially (but not only) those identified in the preceding section (see figure 8.4). At its most simple it takes the opposite approach to traditional production management, which is to try and hide problems; instead it actively seeks them out through systematic analysis and constant monitoring and re-appraisal in the quest for continuous improvement. But JIT is not just about finding problems: it is about mobilizing efforts towards solving them: '. . . the term "expose" is not concerned with identification of problems but with bringing them to attention in a manner that forces something to be done about them.'[10]

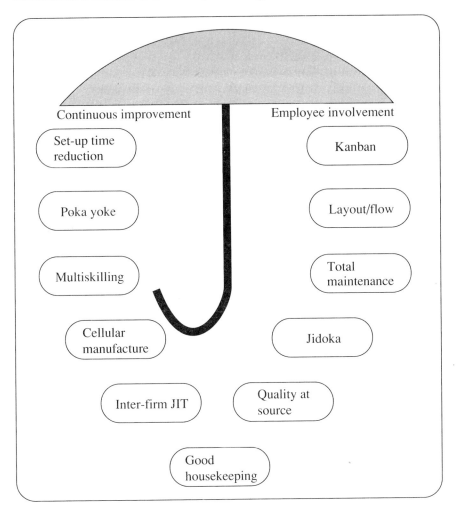

Figure 8.4 The JIT umbrella

Central to any JIT programme are the ideas of:

1 *Continuous improvement* ('kaizen' in Japanese) This is the key underlying value which maintains the momentum in a JIT programme and reflects the view that there is no end to the quest for excellence. In practice this means a regular cycle of problem identification, solution and evaluation, and the view that the

problem is never fully solved – there is always a way of improving further on it. 'Best is the enemy of better' is one typical motto to reinforce this approach. Its effect is to provide long-term support to the principle of solving problems by gradual erosion, rather like water wearing away stone. If we consider the kind of challenge posed to the car industry in terms of set-up time reduction, we can see 'kaizen' in operation. Here the initial problem was how to reduce set-up times of days to hours. The problem did not have a single solution, nor did it have an instant answer. Instead, a repeated sequence of analysis, suggestion, implementation and review took place, leading to ideas which individually shaved seconds or minutes off the set-up time. But the continuous improvement principle meant that these regular attacks on the problem, over a 30-year period, reduced set-up times from days to minutes. 'Eat your elephant a spoonful at a time' is another appropriate motto here!

2   *Common ownership of problems.* In a system in which everyone is someone's customer, pulling products through the factory, questions of problem ownership are crucial to the success of JIT. Passing the buck is passing on a problem rather than exposing it and trying to solve it. Accepting that it is not the responsibility of someone else but that it is a shared problem builds support for effective continuous improvement. This can be seen in the case of quality, for example. In a traditional line if there is a quality problem with defective parts or mistakes being made, these are often passed along – 'It's not my job to fix this.' The effect is to compound the orginal problem, building inventory which will have to be reworked, hiding the original cause.

3   *Participation.* Effective problem-solving requires the mobilization of whatever resources can be brought to bear to provide creative ideas for possible solutions. By extending ownership of the problem and involvement in the problem-solving process, firms can enhance their effective problem-solving capacity. To put this in perspective, the history of suggestions made by the workforce in the Toyota Corporation is set out in table 8.2: the management received around 1.6 million suggestions for improvements during 1983 and was able to implement the vast majority of them. This clearly goes well beyond a simple suggestion scheme and puts participative problem-solving at the heart of the company culture.

The basis for this is the belief that, contrary to Taylor's theory, specialist functions do not have a monopoly on good ideas. Indeed, as Shingo and others point out, it may well be that the person best able to comment on how to improve a machine is one who works closely with it daily – maintaining or operating it. But it is not just a company suggestion scheme – it is an expression of valuing employees for more than their hands. For example, in a recent interview with US and Japanese vehicle manufacturers the question was asked 'How many industrial engineers do you have here?' The US answer was 30 (all professional specialists) but the Japanese answer was 1500 – 'Everyone's an engineer here.' Clearly, the firm with the higher number of such engineers has, statis-tically, a better chance of coming up with creative solutions. These need not be world-shattering in themselves: the principle of 'kaizen' argues that a small step forward every day is better than a giant leap every few years.

Year	No. of suggestions	Percentage implemented
1960	9 000	39
1973	247 000	70
1976	380 000	83
1979	575 861	91
1980	859 039	94
1981	1 412 565	94
1982	1 905 642	95
1983	1 655 858	96

Source: cited in Hoffman and Kaplinsky, 1988

Table 8.2   Suggestions made and adopted within Toyota

Core JIT principles

A key feature of JIT is that it represents an alternative to the traditional Fordist models for mass production, adapted for small-volume production and stressing minimum waste and high quality. Many of the principles which underpin all the JIT techniques – for example, closer integration of tasks and roles for workers and less specialization – can be seen as alternatives to models which are suited to mass production but are less suited to high-variety, low-volume activity.

Most commentators indicate that there are a handful of 'core' elements to a JIT system, beyond which there is a battery of techniques which may or may not be appropriate to particular circumstances. For example, Kaplinsky and Hoffman[11] highlight seven key elements:

- demand-driven production
- flexibility in product and process
- multi-skill and multi-task work
- just-in-time production (minimum inventory)
- zero-defect policies
- giving responsibility back to the worker
- worker involvement in technical improvements

Voss and Ozaki-Ward[12] suggest that there are at least 128 techniques described in the literature or practised in firms, but these can be split into a handful of core techniques and a much larger number of peripheral methods which represent 'tools' and 'enablers' for JIT. They group the core techniques into three classes:

- those concerned with improving flow – including layout, material handling, cellular manufacture, group technology, preventative maintenance, focus on process balance and the use of multiple small machines rather than sophisticated large ones
- those concerned with improving flexibility – including very small batch operation, reduced set-up times, workforce flexibility in skills and working practices and spare physical capacity
- developing the supply chain (sometimes called JIT-2, reflecting the distinction between in-company and between-company JIT) – these include quality policies, changing supplier relationships and smoothing of production rates

It is important to remember that these techniques work together rather than in isolation, attacking the key problems of operating JIT. In the next few pages we will look at some of the most important techniques which form a key part of most JIT implementation.

Production layout

Continuous flow manufacture is clearly the most effective way of producing, but this option is open to only a small percentage of manufacturers. Attempts have been made – notably in the automobile industry – to obtain approximations to such continuous flow, but in general manufacturing is an interrupted and discontinuous process. JIT seeks to challenge this by intervening wherever necessary to try to obtain smooth flow through the plant. To use the river analogy again, rivers with rocks and obstructions which are hidden on their beds flow turbulently and with high resistance, slowing them down. Rivers flow fastest and smoothest when there are no obstacles. Therefore JIT aims to achieve smooth flow, by dealing with problems rather than living with them. A key part of the process is to change the layout.

In many cases factory layout has not been governed by any systematic principle but rather by evolution: as products change and machinery is replaced, so the patterns of flow become more and more convoluted. It is sometimes instructive to track a product through the stages of its manufacture to find how often it goes in and out of stores, the route it takes while being transported round the factory and the time it spends in transport. It is often possible to identify parts which take days and travel miles within workshops measured only in feet, and subject to processing limited only to hours. This is clearly a source of extensive waste.

One common contributor to this problem is the idea of functional layout. In this model machines associated with a particular process, such as grinding, will be grouped together. The justification for such a layout is that it brings particular operations under control, but it can have negative implications for the wider effectiveness of production. Material flow might well follow the chaotic pattern shown in figure 8.5. Problems which may arise as a result of this arrangement include:

- problems in coordination and scheduling
- build-up of WIP
- multiple handling of material
- transportation waste
- problems in tracking and control
- long lead times
- difficulty in tracing quality problem sources
- minimal flow and low opportunity for standardization

This arrangement also limits communications between people and thus precludes both the identification and solution of problems. It also has the effect of reinforcing the barriers which grow up between different functions: the typical 'us versus them' attitude which can lead to too much emphasis being placed on local efficiency and operation, and not enough on the broader organizational picture and the effectiveness of the business as a whole.

An alternative approach is to group machinery and people by product in what is often called group technology (figure 8.6). Here facilities required for making a family of similar products are grouped together and flow is through production cells or modules, rather than via functions. In this kind of arrangement, processes are linked much more closely and the

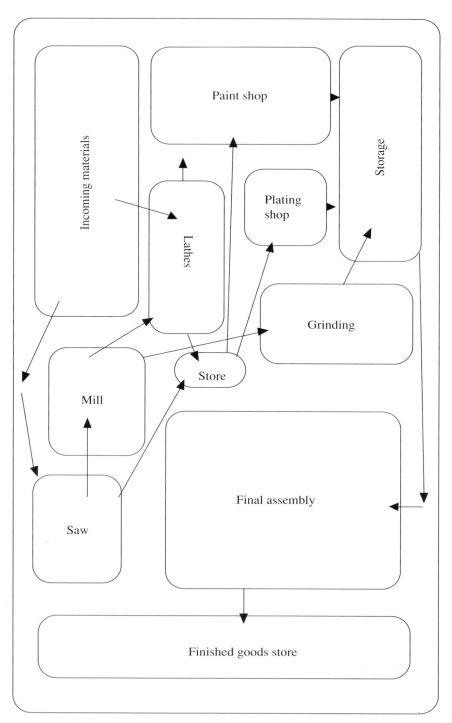

Figure 8.5   Typical flow of material in a process layout

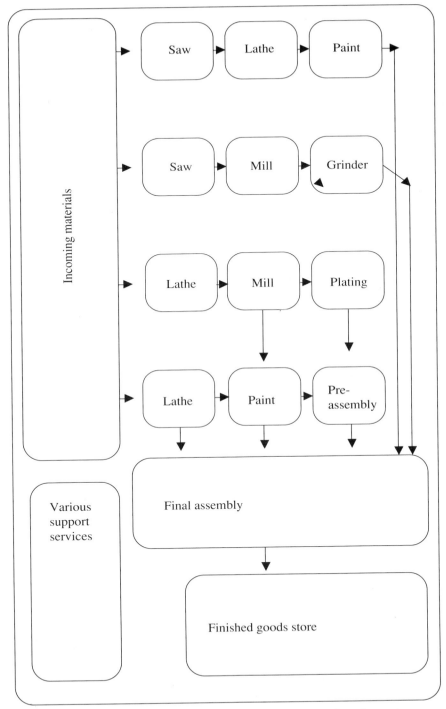

Figure 8.6   Typical flow of material in a product layout

logic behind material flow is much clearer. Roles within such layouts are changed, relating to a particular product rather than to a process. In some cases the idea of a 'factory within a factory' is used, bringing all the relevant functions needed to support the manufacture of the product (stores, maintenance, quality control, and so on) into the mini-factory.

Product-type layouts also imply alternative working patterns, moving towards more team work, collective responsibility and payment by group results rather than emphasizing the individual. Here again we can see the challenge to 'accepted' models of manufacturing organization which date back to Taylor and Ford and which essentially use individual incentives, division of labour and extensive planning and task allocation. By contrast, product layouts based on cells hark back to an earlier era in which craftsmen worked on producing the complete article rather than just a part of it – and had a correspondingly high degree of commitment to its quality and to the service which they offered the customer.

The link to other JIT principles is inescapable. Effective cellular manufacture depends to a large extent on multiskilling and flexibility and on workers taking responsibility. Like much of JIT, such models are not Japanese in origin. Much of the credit for establishing the operational principles of group technology, and for 'crusading' for its introduction, must go to Burbidge.[13] Models such as Gigli's for 'period batch control' were used in the Second World War for, amongst other products, the manufacture of Spitfire aircraft. (This is an approach in which a cell of workers is responsible within the framework of a broad plan for deciding when and how products are manufactured, achieving considerable flexibility in the process).[14] Other pioneering exercises in group working and alternative approaches to layout include considerable work at the Tavistock Institute in London and projects in Scandinavia, in particular the Volvo experiments of the 1960s and their more recent programmes such as the new plant at Uddevalla.

One key attraction of the concept of cellular layout is that it allows for flexibility in both volume and variety of products being made. By emphasizing the general-purpose machinery available within each cell, rather than sophisticated flexible machines available only at one location, it becomes possible to switch production to cope with high variety or high volume. Thus, when demand for product A is high, all cells can be configured to make it; while when it is for B, C, D, E and F, each cell can be configured to make one variant only. There may be under-utilization of some machines in this arrangement, but this is offset by better flow and reduced inventory (see figure 8.7).

Once again, the idea of continuous improvement is central to success. There is no ideal layout, only problems with the existing one which can continuously be improved upon. In this way many ideas for reducing

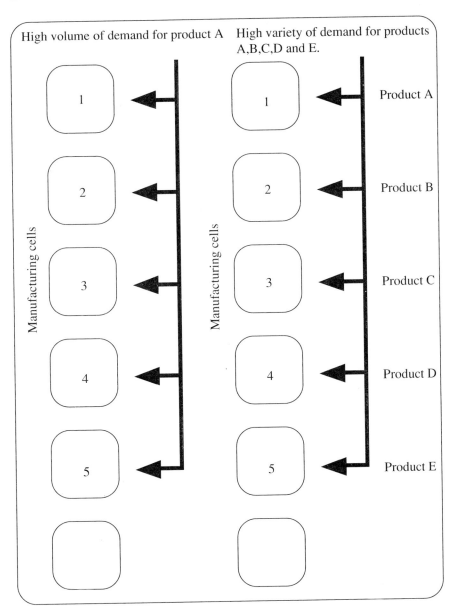

Figure 8.7   Volume and variety in product layouts

transport wastes or delays and bottlenecks with particular kinds of production can be identifed.

Another important contributor to smooth flow is the idea of U-shaped production lines. These represent a regrouping of machines or assembly stations to minimize the movements which workers need to make to carry out operations or to access stores. A direct benefit of such arrangements is the reduction of space required for operations, often representing a saving on floor space of 30–40 per cent.

U-shaped lines also have the benefit of improving communications, bringing workers into more direct face-to-face contact: as distinct from the traditional linear arrangement in which there is a minimum of opportunity for interaction. The benefits of communication emerge in suggestions for process and product improvements and in generally better morale.

Smooth flow can also be aided by careful attention to materials handling. Rather than trying to compromise and accept layout problems, and then deal with them through the use of fork-lift trucks and manual movement, much can be achieved by reconfiguration. Once this has been done, the next step can be to couple lines more tightly together through the use of conveyors and special-purpose fixtures.

The elimination of bottlenecks and queues is clearly of central importance in securing smooth flow and this places emphasis on preventative maintenance to avoid machine breakdown. It also argues for the use of spare capacity and of multiple machines rather than concentrating resources on single complex machines, since if the former break down there is the option of alternative routing, minimum disruption and the avoidance of bottlenecks.

Kanban

The concept of 'Kanban' is derived from the Japanese word for 'card' or 'visible record', and is used, essentially, as a way of enforcing the discipline of a pull approach to inventory use throughout the factory. It is one of the best known elements of JIT, with the result that many managers assume that JIT is simply the application of a pull system in inventory management. Its origins are, once again, to be found within Toyota and are attributed to Taichi Ohno, the founder of the company.[15] In the 1950s he was attracted to the ideas of applying principles used in supermarket shopping to manufacturing management, and in particular was struck by the way in which the customer takes just what he or she needs in the right quantity at the right time. The supermarket only replenishes what has been used up. He began experimenting with such ideas, but the problem which held him back was how to impose the discipline of the supermarket system onto manufacturing operations.

Kanban was born as a response to this problem. A kanban is a record of some form – a card, a tag or a bar code – which holds information such as the part number, the quantity of parts, their source and destination, and so on. In this it is very similar to a batch card used for production control in any factory. However, its use is critical. For example, products on some supermarket shelves carry a kanban, and these are then collected at the checkout (by bar code scanning, for example) to provide an exact indication to the stores manager of what needs to be replenished on the shelves. Thus the kanbans act as a signal or trigger for pulling just enough inventory from the warehouse and into the shop.

We could go even further still and, instead of a warehouse or suppliers delivering finished goods, imagine that we actually connect the factory making them to this system. The instructions for how much to make (assuming a highly flexible factory) come from the kanban cards, which indicate a need. As each new product is made, its production card can be swapped for a need card – and the cycle can be repeated.

The advantages of this system are considerable. It imposes a simple but powerful discipline on the whole manufacturing and sales chain so that inventories are kept to a minimum: following JIT principles nothing is made until a need is expressed for it and goods arrive just in time to be sold in the store.

This principle can be applied equally well within the factory, relating different stages in the process with their needs and with production. In an ideal sense this would look like figure 8.8, in which kanban cards are exchanged along the line and there is instant response.

In practice this is not always possible: machines have finite lead times and some buffer stores are necessary to keep things flowing. In effect, referring back to our example above, these interim stores act as mini-warehouses. But they do represent waste and so, in the long term, should be considered targets for inventory reduction and eventual elimination. Therefore figure 8.9 portrays a more practical starting point, with the goal of gradually linking production more tightly by elimination of kanban stages.

Basic 'rules' for effective operation of a kanban system include the following:

- people from a downstream process only obtain parts from the upstream process via a kanban card signal

- people upstream only produce according to the information on the kanban card

- no kanban = no production or issue of materials

- if there is a problem with any parts going into the parts container for downstream use, the process should be halted until the defect is sorted out – no defective parts should be produced or issued

- over time, reduce the number of kanban cards to link processes more closely together

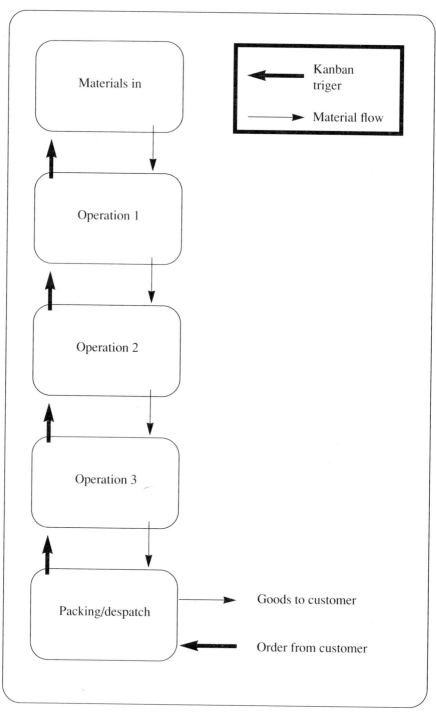

Figure 8.8 Ideal kanban flow

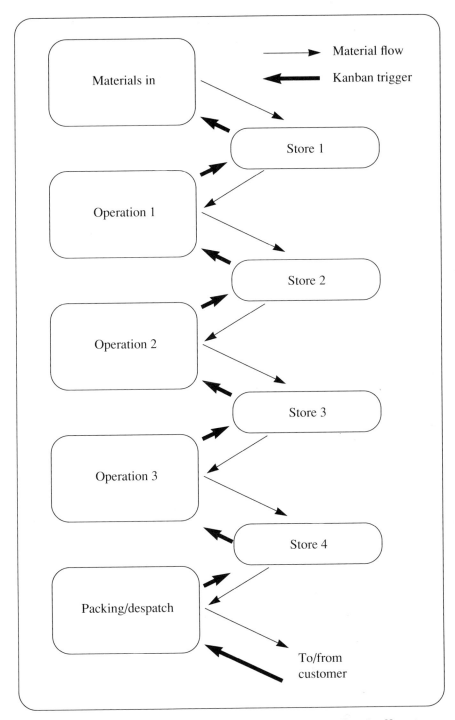

Figure 8.9   Typical practical kanban system, including buffer stores

These rules effectively make kanban not only a device for production scheduling and control (since information is held on the cards and can be used to track things through) but also an approach, to process improvement since the elimination of kanban stages is a goal. In the Toyota system, for example, there are standard containers, each of which carry a kanban card. The foreman can control the amount of inventory in the system by removing kanban from the system.

In practice, kanban may take many forms. One important distinction is between single- and two-card kanban. Where production is sequential and there is a direct link between stages, the system can operate with a single card that indicates a signal to produce and to move material to the next stage. But where there is a more complex arrangement – for example, an assembly operation that is fed from several different production locations – then separate cards triggering movement and production may be used, and not necessarily at the same time.

Again, there may be cases in which it is more economical to run a small batch at a time rather than produce in batches of one. Here the kanban cards can be used to pull materials out of store until sufficient have been used to to trigger production of a new batch (rather like a reorder point form in stock control). Similarly, some assembly operations require materials from a number of different locations; rather than issue one card per operation, the relevant materials can be kitted up in short-term store and moved on receipt of a single kanban card – as in the example shown in figure 8.10.

Some kanban arrangements do not rely on cards but make the container itself the kanban, carrying both information and parts. For example, in the example below the firm use a trolley which serves both as a parts carrier and as a kanban signal in its own right. When the empty trolley appears in stores, it is a message that the next stage of production is ready for a new set of parts.

Another variant is the use of kanban squares, areas of the plant marked off in squares between process operations. Here the pull discipline is imposed by not allowing inventory to accumulate between process stages. If the upstream process is ready but the downstream cannot handle it yet, there is no point in producing more.

One major problem of kanban is that in its ideal form it only produces to order. Therefore sudden surges in demand may result in delays and poor customer service, while large orders which are suddenly changed may result in excess inventory being made. Fluctuations can be handled by pushing more inventory into the system, but this goes against the whole principle of JIT. Instead, the production process and the implementation of kanban need to be adapted to cope. Once again, there

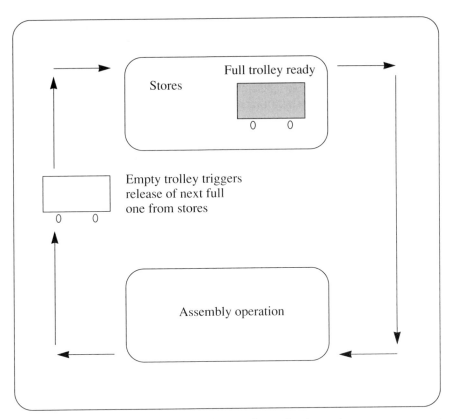

Figure 8.10   **Kanban trolleys and containers for inventory management**

is a need for other components of JIT to operate to support effective kanban; for example, layouts and line balancing should be adjusted to allow
for what Suzaki calls 'levelled/mixed production'.[16] The reduction of
set-up time is critical to ensuring responsiveness and thus reducing the
amount of buffer storage needed. Preventative maintenance minimizes
the uncertainties due to machine breakdowns.

Voss and Ozaki-Ward[17] points out the central importance of a master
schedule to effective operation in JIT mode, and this is particularly
critical for kanban. A number of firms (even in Japan) find kanban un-
suitable for highly complex activity, while others are increasingly
looking at hybrid systems which use sophisticated planning packages
such as MRP2 as a way of generating a master schedule and then
operating kanban within that plan.

Reduction of set-up time

The reduction of set-up times is crucial to JIT, since it offers increases in
flexibility, shorter lead times and reduced inventories. It also offers firms

the chance to compete effectively in markets characterized by demands for greater product variety or more rapid product innovation.

Set-up time consists of four basic components which apply to all types of machine. These are:

- preparation and finishing; getting parts, fixtures and tools ready, delivering them to the machine, and the reverse of the process, taking the old ones away (together with any necessary cleaning or maintenance) – this component represents about 30 per cent of set-up time

- mounting and removal of tools and fittings – this component represents around 5 per cent of set-up time

- measuring, calibrating and adjustment so as to ensure the correct positioning, speed, temperature or other conditions are fulfilled. this component accounts for around 15 per cent of set-up time

- trial runs and adjustment, to ensure that the machine is now set-up correctly – this accounts for around 50 per cent of set-up time (clearly, the better the adjustment, the less time is required for this stage)

Reducing set-up times was one of the major breakthroughs of the orginal JIT efforts within Toyota and owes much to the work of Shigeo Shingo[18]. His approach, which has become known as the Single Minute Exchange of Die (SMED) system led to massive reductions in set-up times across a range of equipment, not only in the large pressworking machinery area but across the manufacturing spectrum. Some idea of the reductions which have been achieved using these principles can be gained from a study of figure 8.11.

In essence, Shingo's system evolved through a classical piece of work study, watching how changeovers were effected and constantly trying to improve on them. His observations took place in the 1950s in a variety of factories, and the system was refined during the 1960s in his work with Toyota. The process made extensive use of the principles outlined above, of involving all the workforce and never accepting that the problem had been finally solved. Indeed, the momentum was maintained at times by senior management setting apparently impossible goals. For example, as Shingo comments, '. . . in 1969 Toyota wanted the setup time of a 1000 tonne press – which had already been reduced from four hours to an hour and a half – further reduced to three minutes!'[19]

The consequence of this is that the system is basically simple and can be applied to any changeover operation. It relies on observation, analysis and creative problem-solving. There are four basic steps:

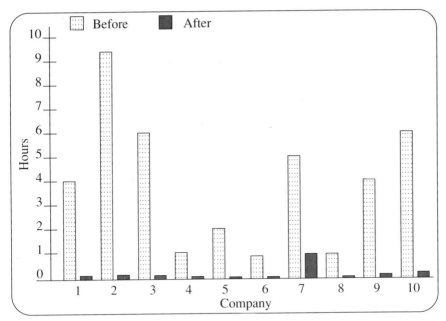

Figure 8.11   Set-up time reductions. 1, Toyota, 1000 tonne press, 1945–71; 2, Yanmar Diesel, machining line, 1975–80; 3, Mazda, ring gear cutter, 1976–80; 4, Hitachi, diecasting machine, 1976–83; 5, US chain saw manufacturer, punch press, 1982–3; 6, US electrical appliance manufacturer, 45 ton press, 1982–3; 7, Venezuelan car components manufacturer, sintering press, 1984–5; 8, Bridgestone Tire, vulcanizing process; 9, Kyoei Kogyo, bending and piercing press; 10, Kubota Industries, machining line.
Source: **complied from Suzaki, Shingo and the author's research**

1   Separate out the work that has to be done while the machine is stopped (called internal set-up) from that which can be done away from the machine while it is still operating (external set-up). For example, typical external set-up activities include the preparation of dies and fixtures and the actual movement of dies in and out of stores and to and from the machine in question. Typical internal tasks are attaching/detaching dies, adjustment and testing. Shingo estimates that between 30 per cent and 50 per cent of set-up time can be saved by doing external set-ups rather than including these operations as part of internal set-up.

2   Reduce the internal set-up time by doing more of the set-up externally. This can be done, for example, through the use of pre-set dies or special easy-fit fixtures holding the new dies. External set-up times can themselves be reduced by colour coding, easy access to stores of tools, special transport equipment,

and so on – all designed to minimize the time taken for moving and finding.

3   Reduce the internal set-up time by simplifying adjustments and attachments, developing special easy-fit connections, and devoting extra resources to it at the critical time (making it a team effort, with several people leaving their 'normal' jobs to help out with the changeover, and so on). Adjustment of dies once in position often accounts for up to 50 per cent of the total internal set-up, so finding ways of reducing or eliminating the need for adjustment (for example, through the use of locator pins or grooves) is a rich source of opportunity. Another example, quoted in Suzaki, is the pre-heating of dies for injection moulding machinery so that the machine is able to start running much earlier.[20] Another important way of reducing internal set-up is the use of parallel operations, where extra assistance is provided at the critical time of changeover and the time is reduced by means of several people working in parallel. This not only improves things by having extra pairs of hands but it cuts out time which a single operator would spend fetching, carrying and fitting. Suzaki gives a useful illustration of a set-up carried out in parallel, in which the simple addition of an assistant for some of the key stages in a changeover cut the internal set-up time for a press from 57 to 10 minutes.[21]

4   Reduce the total time for both internal and external set-up. Implicit in this is the concept of continuous improvement, of constantly monitoring, analysing and developing new attacks on the problem.

Changing over set-up is thus a precision, team-based activity in which everything is to hand, special-purpose tools and fixtures are used, and everyone knows what he or she has to do. A very close and effective analogy can be seen in motor racing when a car makes a pit-stop for tyres to be changed. This is also accomplished in seconds, as a result of a very similar process.

The reduction of set-up time demonstrates a number of important JIT principles, especially that of continuous improvement. Opportunities can be identified through 'watching with new eyes'; taking a closer look (perhaps with the use of a video) and discussing ways of separating out activities. Analysis, using various techniques (such as PERT networks to determine critical paths) and creative problem-solving using the whole group through techniques such as brainstorming can generate many potential improvements. Over time the pattern of experimenting and repeated attacks on the problem means that the set-up time gradually falls. Recording progress, for example, on a downward-sloping graph, so that all can see is an important way of continuing to motivate people in the process. There is the added bonus that ideas developed for one set-up time reduction may well be transferable to other operations or product areas, and demonstrations of 'before and after' set-up time reductions can also motivate other groups of workers and help to boost morale by sharing success.

Multifunction workers

JIT requires a high degree of flexibility in working practices to ensure that there are the necessary skills where and when they are needed – just in time for them to be used. This implies alternative patterns of work organization, payment systems, and so on along the lines suggested in chapter 5. Above all, it requires that workers are trained to be multifunctional, able to carry out a variety of tasks and to switch quickly between them.

These do not necessarily have to be high-level skills. For example, statistical process control is something which can easily be taught and which makes everyone a quality inspector. This does not mean the elimination of the Quality department, but it does become a place in which emphasis can be placed on the strategic development of quality, because there is less 'firefighting' for the highly trained experts to spend their time on. The same principle applies to maintenance: everyone can contribute to simple preventative maintenance and good housekeeping, with the result that experts can concentrate on planned and preventive activity rather than on crisis management. Other activities, such as product and process innovation, and production control, can similarly become part of an operator's job.

JIT requires workers to be seen as resources who can be developed and trained in order to provide flexibility. Once again, this fits with the other pieces of the JIT jigsaw: for example, only such multiskilled workers can make group technology cells work. The role of training is central in all of this: without investments in regular development and updating of skills, the necessary flexible base will not be available, and the opportunities for implementing many of the proposed solutions to continuous improvement problems will be curtailed.

Developing such multiskilled, flexible workers also requires alternative arrangements for work experience. Rather than tie people to a particular task, JIT emphasizes principles such as job rotation and teamwork, which provide greater variety, better motivation and also the opportunity to 'see things with new eyes' – an important feature of continuous improvement. Clearly, such flexibility in deployment will depend on the degree to which the threat of redundancy or of worsening conditions of employment can be allayed – something which is achieved in large Japanese firms through the lifetime employment system.

Appropriate automation

A prerequisite for smooth flow is that machines operate effectively. As we saw in earlier chapters, IT has much to contribute in improving the reliability, accuracy, speed and overall potential of equipment, even in retrofit mode. Set-ups can be reduced through the use of programmable controllers and, on a larger scale, more integrated systems can offer both flexibility and overall improvements in productivity. There is no conflict between automation and JIT, but only a requirement that whatever is done is in relation to the total system, and not simply introduced to improve a single operation without considering its effect on the rest of the system.

For example, installing a powerful reprogrammable robot on a line which can only operate at a fixed speed will be useless – a much simpler solution would be a special-purpose pick and place device. If only one product is to be made this does not even need to be reprogrammable.

Simple machinery and low cost automation – often retrofitted – is an important component of JIT. As the process of continuous improvement advances, so the simple machinery can gradually be substituted for more powerful equipment, but always remaining appropriate to the requirements of the whole system and the ability of the workforce to support it.

The principle of 'jidoka' – or autonomation – reflects this view. It involves adding some form of 'intelligence' to equipment so that it can operate to support or extend operator capabilities. The goal is to eliminate the need for the operator, so as to free him or her for more creative use of his or her skills and intelligence. Machine minding is a typical example in which fitting some kind of intelligence so that the machine can detect problems and inform operators, or fail safe, elimi-nates the need for wasteful activity. Applications of jidoka can be very simple; for example, fitting limit switches or warning lights to existing equipment.

Jidoka also works in conjunction with operators where they are part of a larger machine or process. For example, intelligence is built into the assembly lines in the car industry through the concept of the linestop and the use of 'Andon' lights. Here, if a problem emerges – for example, a defective incoming component – then there is no point in continuing manufacturing until the problem is solved. Andon (which means 'lantern' in Japanese) warnings can be switched on by any employee to notify problems and focus attention quickly on the problem. In the extreme case operators can stop the entire production line. Different-coloured lights, or buzzers and lights in combination, can be used to signify degrees of difficulty being encountered.

The key here is the need to mobilize a sense of ownership of problems, and a participative approach to solving them. Unless operators see it as their problem, they will not stop the line or switch on their lights – just as they will not come forward with suggestions for improvement. Equally, unless top management express their commitment by trusting operators with the full authority to stop the line, then the whole principle is worthless.

Poka yoke: foolproof mechanisms

An extension of the continuous improvement approach applied to machinery and processes is that of foolproofing them – making sure that mistakes cannot be made – through various simple devices. Many of these result from employee suggestions and individually they may make little difference. But grouped together they can make a considerable difference to reducing set-up times or increasing quality. 'Poka yoke' ideas might include getting engineering to make up a special fixture on which it is impossible to locate a part wrongly, or something to eliminate the need for judgement in adjustment of a press die. Templates, limit switches, colour coding and so on, are all typical examples of poka yoke.

Associated with this is the idea of simplification – of constantly challenging the complexity of existing solutions in search of simpler ways of doing the same job. In many cases the simpler methods are more reliable and less wasteful.

Total quality control

We will discuss this in detail in the next chapter, but it is important to stress here that another essential feature of JIT is the availability of defect-free parts at all stages in manufacture. In turn this requires an awareness of the quality problem, which is shared and owned by everyone – and for which everyone carries responsibility. This gives rise to the principle of 'quality at source' as distinct from quality control as a separate function. In this arrangement everyone is responsible for checking quality, and for not passing on anything less than 100 per cent perfect to the next 'customer'.

One way in which quality at source is obtained is through the use of statistical process control (SPC), which also imposes an important discipline on the entire manufacturing process, identifying key sources of error and focusing problem-solving activity on them. SPC means that instead of controlling quality by inspection or checking at the end of a process, it is continuously checked in line at every stage of the process.

SPC is a simple approach (described in detail in the next chapter) which allows operators to identify quickly when their processes are out of line, and to stop developing quality problems before their impact becomes serious – by making scrap, and so on.

Continuous improvement groups

A number of approaches are used in order to implement continuous improvement as a company-wide philosophy. They share the idea of regular, team-based sessions, which take place in the company's time, in which groups of workers meet and, through the use of aids to problem analysis and solution, arrive at solutions which can be put to management for further action. These have a variety of names, probably the most well known of which is the quality circle (which will be discussed in detail in the following chapter). It is important to recognize that the remit of these groups is much wider than quality: it extends to elimination of waste in all areas, reduction of set-up times, improvements in maintenance practice, poka yoke ideas and so on. Much of the success in generating a high volume of practical suggestions results from the operation of such groups rather than through individual-oriented suggestion schemes.

Visual control or management by sight

Part of the simplification of JIT is the attempt to make everything visible rather than hidden, and to reveal problems quickly so that they can be dealt with. Various techniques exist of making things more visible: we have already seen the concept of andon lights to indicate where and when trouble is occurring, and of SPC, which involves the use of easily visible charts on which it is clear when a process is slipping out of line. Other indicators include progress indicator boards which highlight how well the company is doing in reducing inventory, quality problems or set-up times. Colour coding is also widely used.

Preventative maintenance

Most problems with maintenance are of the 'fire-fighting' type and involve dealing with unexpected crises and breakdowns. A more desirable approach is clearly one of prevention, but this simple notion is rarely applied in many Western factories. However, JIT depends upon reliable machinery for ensuring smooth flow and so considerable priority has been given to developing ways of guaranteeing effective maintenance.

Much as the process of set-up time reduction evolved out of classical work study and continuous improvement work, so total preventative maintenance (TPM) has grown into a valuable approach. It is recognized that many of the basic causes of problems can be dealt with through the use of simple approaches such as; regular lubrication, and inspections which can be carried out by workers rather than experts, whose specialist knowledge can be more usefully deployed elsewhere. Analysis and discussion of continuous improvement activity and the causes of failure also makes a major contribution, identifying key causes of failure and devising poka yoke solutions to help avoid common mistakes. An important feature of this approach is to try and attack the right problem at the root of the maintenance difficulty, rather than the most obvious one which may in fact simply be a symptom. Techniques such as fishbone charts and critical examination are valuable here.

Although much can be done through the use of operators as multi-skilled workers doing first-line mainteance, there is still a need for gradual upgrading and improvement of machinery and systems to reduce the likelihood of breakdown further.

Suzaki cites a four-phase approach to machine improvement, originally developed by the Japan Institute of Plant Maintenance.[22] Here the aim is to reduce the incidence or likelihood of machine breakdown by systematically attacking the main problems, gradually building up to a high level of preventative maintenance.

In phase one breakdowns due to what are termed 'forced deterioration' are focused upon. Forced deterioration is the result of neglect and bad maintenance – lack of lubrication or inspection, for example. Much machine downtime can be eliminated by addressing these problems systematically.

Phase two addresses what is termed 'natural deterioration', in which problems arise as a result of normal operating life of the machines. These can be tackled by regular attention to operating features such as the tightening of bolts.

In phase three the machines themselves are adapted and developed to include poka yoke ideas and some parts are replaced or redesigned to extend machine life. The final phase extends this active maintenance and development to full condition monitoring, in which the machine's condition is constantly monitored and deviations are dealt with – essentially a similar process to SPC, described earlier.

One final point about TPM is the emphasis on the total system approach. Just as with total quality management, so effective maintenance depends upon generating the environment in which everyone feels that they *own* the problem of maintenance and act to help solve it.

Levelled or mixed production

We have already seen the potential advantages of being able to offer small-batch production in terms of meeting market demands for variety and volume shifts. But to produce effectively we also need some way of levelling out the pattern of demand so that the output rate becomes standard, and the remaining operations can be synchronised.. To achieve this requires close co-operation and communication between production and marketing. The goal is to achieve a fixed level of production which will also satisfy the market.

Traditionally this can be done by holding a high level of finished goods, so that production can plan smoothly to make for stock and marketing can sell from stock. The challenge in JIT is to achieve this without that high level of finished goods. In practice this can be achieved by agreeing production schedules for short periods. Within these periods what is agreed is fixed and production gears itself to this invariable schedule. By making the periods shorter and shorter – and by developing a rapidly responsive and agile factory – it is possible to achieve a balance between manufacturing and marketing. Not surprisingly, the balancing act is complex and many firms use computers in order to make JIT scheduling possible.

Housekeeping

Another very simple but powerful tool which contributes to improved flow, problem recognition and better maintenance is the systematic application of some basic principles of good factory 'housekeeping'. In particular, JIT emphasizes what in Japan are called the four S's; 'seiso' (clarity), 'seiketsu' (cleanliness), 'seiri' (orderliness) and 'seiton' (tidiness).

Supply chain management (JIT2)

As we saw in the supermarket example, the concepts of JIT can be extended beyond the factory or shop gates and back into the supply chain. In theory this is simple, but in practice achieving JIT delivery of the right number and quality of components at the right time is often problematic. Indeed, achieving JIT2 requires a shift in the basis of relationships: this is only partially about physical links and has a great deal to do with management attitudes, We will pick up these themes in chapter 10, but for now it is important to stress the need for suppliers to become involved as partners within a long-term relationship.

Design for manufacture (DFM)

The central importance of this concept was outlined in chapter 7 and it is clear that Japanese firms have begun to apply team-based cross-functional approaches very effectively. DFM is yet another example of the integrating effect of JIT. For example, simplification and standardization, reduction of process stages, improvement of quality, reduction of inventory or materials requirements are all things which can be achieved through good design for manufacture, and improved further if the feedback loop is closed.

Standardization

The intention of standardization is not to make every product the same but to make the process controllable and predictable by reducing variance. In practice, this can be achieved in a variety of ways, including reducing the number of parts and the routings which they follow, modular fixturing, containers and handling, uniform output rates (tying everything to the speed of the production line), Statistical Process Control (which aims to reduce the variance due to quality problems) and automation. In addition there is the emphasis on using the factory as a laboratory for constantly improving product and process design by rounding edges and designing for manufacture.

8.4   DIFFUSION OF JUST-IN-TIME IDEAS

JIT is a low-cost approach which, as can be seen from the above, offers a wide range of potential gains. Its diffusion in Japan is widespread amongst larger firms, and in recent years has begun to accelerate in other countries. For example, Schonberger documents its growing application in North America, listing at least 100 companies on his 'honor roll', and Wildemann, in the Federal Republic of Germany, has been working with over 100 firms that are introducing JIT.[23] A number of applications have also begun to emerge in developing countries, especially amongst the newly industrializing nations of the Far East and Latin America.[24]

JIT took some 30 years to evolve in Japan and even now there is no clear indication of the extent to which all the above practices are in use. Amongst Western nations progress has been slower, and emphasis has often been placed on some techniques at the expense of others, and on piecemeal rather than systematic implementation.

For example, Voss and Robinson report on a survey amongst UK manufacturers in 1986, which drew responses from 132 firms.[25] Of these some 57 per cent had implemented, or were planning to implement, some form of JIT, although only 16 per cent had actually put a formal programme for investigating and implementing JIT into practice. The size and sectoral breakdown of the sample underlines the particular interest which has been shown in the electronics and engineering industries for JIT, but also highlights that there is no major barrier preventing smaller firms from becoming involved.

The main techniques applied are indicated in table 8.3, from which it can be seen that emphasis is particularly placed upon inventory reduction and increasing workforce flexibility. The *benefits* reported by these firms include the following (in order of importance):

● inventory (WIP) reduction

● increased flexibility

- reduction in the parts and raw materials requirement
- increased quality
- increased productivity
- reduced space requirement
- lower overheads

Techniques used	Percentage
Flexible workforce	80
WIP reduction	67.1
Product simplification	60
Preventive maintenance	60
SPC	58.6
Set-up time reduction	54.3
Continuous improvement groups	54.3
JIT purchasing	51.4
Work team/quality circles	50
Standard containers	44.3
Modules/cells	44.3
Zero defects	34.3
Mixed modelling (using the same line to assemble different products)	31.4
Smoothed line build rate	25.7
Parallel lines	22.9
U-shaped lines	22.9
Kanban	11.4

Table 8.3   Diffusion of JIT techniques

Of particular interest in this connection is the fact that the benefits which firms actually obtained were often derived from those techniques of which they had expected least, or which had been least explored. The most beneficial techniques were (again, in order):

- zero defects
- WIP reduction
- kanban
- JIT purchasing

- U-shaped lines

- work team quality circles

- modular/cellular manufacturing

- set-up time reduction

- workforce flexibility

- parallel lines

Another UK-based study was carried out in 1986 for the magazine *Engineering Computers* and the consultants Peat Marwick, Mitchell.[26] This survey of 100 managers revealed the benefits set out in table 8.4:

Factor	Percentage responding
Reduced inventory/WIP	74
General financial benefits	26
Reduced lead time	23
Improved quality	19
Less waste or scrap	16
Improved supplier relationships	12
Reduced space requirements	11
Flexibility and faster response	11
Improved customer service	10
Better control of production	9
Increased worker motivation	5
Increased productivity	5
Less handling required	3
Better manpower utilization	3
Reduced set-up times	3

Table 8.4   Experience of JIT

At the level of the individual firm there are many examples of successful implementation of JIT. Cummins Engines in Scotland[27] reduced WIP inventory levels by 26 per cent, floorspace by 24 per cent, overheads by 18 per cent and response times by 40 per cent. The overall savings averaged out at around $3 million per year. In the Harley–Davidson plant in the US,[28] the inventory of purchased parts and WIP was cut by 60 per cent, scrap by 46 per cent, warranty claims by 36 per cent and overall productivity (in vehicles per employee) increased by 39 per cent.

Two final points are worth highlighting in this section. The first is that many of the benefits due to JIT lie in the area of *intangibles*; benefits which are difficult to quantify or measure, but which nonetheless have a major impact on organizational effectiveness, such as improved motivation or a more supportive culture.

The second is that, as Schonberger has observed, JIT represents a challenge to many of the existing ways of doing things. In applying JIT the embedded 'custom and practice' of 'doing things the same way because we've always done them this way' can be re-examined, and the process of change can have positive effects – a self-healing process.

8.5 OBSTACLES TO THE SUCCESSFUL IMPLEMENTATION OF JIT

Despite the enormous potential attractions of JIT as a low-technology, low-cost, low-risk route to improvement, achieving the promised benefits is not always easy. A survey of 100 firms undertaken by the *Financial Times* in 1986 reported that:

> . . . the bad news is that even among the managers who are involved with JIT, the overwhelming majority are at a very early stage in thinking about workflow, lack a proper understanding of the overall concept and are trying to introduce changes haphazardly and in small piecemeal packages.[29]

Voss and Robinson echo this concern in their survey: '. . . many companies were focussing on easy to implement techniques and were neglecting core JIT techniques which would yield far better payoffs'.[30] Once again we see the use of a powerful technological resource in an incremental way rather than as an integrated technique.

It will be useful to examine some of the major problem issues which act as barriers to more complete and effective implementation of JIT.

Problems of supplier relationships

JIT2 depends on being able to work with suppliers to reduce the overall supplier base and improve the quality and flexibility of those remaining. It has worked well in Japan, where the automobile industry in particular has been able to secure regular just-in-time deliveries to the line, eliminating the need for warehouses and reducing stockholding right along the supply chain. An increasing number of Western firms are also moving in this direction, with the vehicle industry again at the forefront.

However, there are major problems associated with this, many of which reflect on issues which we will discuss further in chapter 10.

For example, the UK consultants Ingersoll Engineers carried out a survey in 1987–8 of just-in-time practices in the supply chain.[31] Of the 200 firms who responded, 80 per cent felt that JIT principles were relevant to their operations, but a much lower percentage had actually done something about it. Of particular concern was the problem of management attitudes in materials procurement; a factor of particular relevance since the average lead time for responding to an order was 16 weeks, made up of ten weeks procurement, four weeks production and two weeks distribution. In other words, most of the lead time gains could be made by improving the supply chain.

But, as Lamming and others argue, improving the supply chain depends on developing a new set of relationships which are based much more on co-operation and mutual development.[32] This includes reducing the supplier base and concentrating on building up excellence in those who remain – often to the extent of providing assistance to improve systems and procedures. Yet, in the Ingersoll survey, each of the responding firms had an average of 300 suppliers, of whom less than half were certificated in fields such as quality. Even more worrying was that less than 7 per cent worked with their suppliers to help them reduce costs and improve quality and delivery performance in the chain.

Inappropriate accounting and measurement systems

Production cost accounting often fails to provide suitable tools with which to measure the full benefits of new approaches. For example, Maclean suggests that few firms actually measure the full costs of holding stocks (which are much more than the physical costs of the inventory itself).[33] These include:

Carrying costs
- interest charges on capital invested in stocks
- storage charges (rent, heat, light, and so on)
- storage staffing, equipment and maintenance
- materials handling
- administration
- insurance, security, losses

Ordering costs
- purchasing, accounting, goods reception
- transport

- set-up and tooling costs for internal orders
- production planning and control of internal orders

Stockout costs

- loss of profit on lost sales
- loss of customers
- production stoppages
- rush orders

Amongst changes which JIT requires in cost accounting are the shift away from standard costing and analysis of variance towards more direct measures which emphasize actual costs and flows through the plant. By the same token, productivity accounting based on individual machine utilization or direct labour hours may be less relevant than systems which look at the overall effectiveness of the plant as a whole; for example, by examining the ratio of total input to total output.

As Wheatley points out, the problem is that '. . . performance measures in most companies penalise improved flow rather than reward it'.[34] Most productivity and efficiency measures usually report smaller batch sizes as being a deterioration in efficiency because they include some penalty for setting-up costs. But maximizing efficiency in these terms does not necessarily equate with improved profitability – as we have seen from considering the costs of inventory, long lead times and poor customer service. Instead, there is a need to shift to alternative measures which reflect the contribution to overall effectiveness more accurately; for example, Voss and Harrison[35] argue that in addition to measures of productivity and cost management other indicators of real performance (such as quality, WIP level, manufacturing lead time and space utilization) should be used.

Payment systems

One feature of JIT is its emphasis on the group as opposed to the individual, stressing flexible working teams within cells. But most payment systems have been devised to reward individuals for results, and for the effort they put in, irrespective of whether or not what they do actually makes a difference to the bottom line. The tradition has been to pay people for working hard; but JIT implies a shift to paying them for other features – such as quality or flexibility – and doing so on a team rather than an individual basis. Not surprisingly, this has been the focus of considerable resistance, particularly from individuals who see it as an erosion of their earning power. For team-based systems to be effective alternative emphasis needs to be placed on:

- developing trust and mutual confidence within the team
- developing equitable payment rates and simple, clearly communicable structures
- involving the workforce in design of the payment system

Lack of vision

Despite the fact that information about JIT has been available for many years in the West and is now backed up by a growing number of successful demonstrations, many firms are still reluctant to try the approach and have developed a repertoire of excuses to justify 'why JIT won't work here'. These include the following:

1   'It's alright for Japanese firms but it won't work over here.' Although there are undoubtedly features of Japanese culture which facilitate the development and operation of JIT, the techniques outlined above are essentially transferable to any manufacturing organization, as the growing diffusion of JIT underlines. Indeed, many of the basic ideas – such as group technology or total quality control – were originally developed in the West and then taken up enthusiastically by Japanese managers.

2   'We couldn't get our suppliers to deliver just-in-time'. Variations on this theme argue that the customer firm is too small to influence its suppliers and the distances between them are too great, so that it is essential to hold buffer stocks, or that general environmental and political uncertainty makes it expedient to operate a broad supplier base. This neglects two key features about JIT. First, JIT2 – the inter-firm relationships – is only one element in the overall JIT approach, and even if the supplier in question offered no opportunity whatsoever for change, there is still much of JIT which can be applied *within* the factory to improve flow, reduce inventory and increase quality. Second, JIT is based on actively seeking out problems and trying to solve them, rather than living with them or compensating for them with excess inventory. Even if the problem is apparently intractable, the approach should still be to seek ways to make continuous improvement.

3   'JIT is only for those involved in repetitive manufacturing'. This excuse, often heard from small-batch producers, indicates a lack of understanding of the breadth of techniques which JIT offers. Although it began life in the vehicle industry and was particularly suited to high-volume repetitive manufacturing, many of the techniques – such as set-up time reduction – are more appropriate for smaller-batch manufacturers. Others, such as total quality control, are valid for all types of manufacturing organization.

Similar excuses about the inappropriateness of JIT for continuous flow producers ('We already operate JIT') or for specialist subcontractors again need to be viewed against the wide range of JIT techniques and the general applicability of some of the core principles.

Industrial relations

JIT involves radical changes in the way in which production has traditionally been organized, and this includes the nature of worker representation and communication with management. In essence JIT begins to replace earlier models, in which trades unions and works councils act as go-betweens, by more direct contact in the context of problem-solving groups, quality circles, and so on. Not surprisingly, this has led to concern by both the organized agencies for worker representation (such as trades unions) and by workers themselves. Significantly, in a UK survey of workplace industrial relations there was very little evidence to suggest that workers were opposed to the introduction of new technology, but considerable concern was expressed about changing working practices and organization.[36]

Costs

Although JIT represents a relatively low-cost investment in comparison with some of the computer-based technologies we looked at in earlier chapters, it is by no means cost-free. Costs arise from a number of factors – development of new equipment, reorganization and relocation, and so on – but the major element is represented by training.

8.6 IMPLEMENTING JUST-IN-TIME

Although some of the benefits may often appear quickly, JIT is emphatically not a 'quick fix', a sticking plaster with which to treat the multiple ills of manufacturing. Instead, it is a continuous process of development and improvement which begins by eliminating many of the inappropriate practices and structures and then begins to develop new and effective alternatives. Such change does not take place overnight – and it relies on an underlying cultural shift in the organization. Unless these new values and norms are taken on board and communicated from the top downwards, then JIT is unlikely to have a great deal of impact, especially in the longer term.

Part of the problem, as we noted above, is that JIT is too often seen as a handful of techniques rather than a pervasive and powerful integrating philosophy. But unless firms appreciate the latter point, then they are making use of JIT in essentially substitution mode and will, at best, achieve only a fraction of the potential benefit. If there is a difference between Japanese and US or European firms in their ability to implement

JIT it is in this understanding of JIT as a total approach, a long-term improvement programme.

Some indication of the kind of approach required can be seen in the way in which Japanese firms introduce JIT into plants which they have taken over. Voss and Ozaki-Ward describe a case of such a takeover which highlights the step-by-step approach, building up from simple improvements in housekeeping towards more complex and powerful techniques. As the Japanese production director is quoted as saying:[37]

> . . . the process of intrduction of JIT must be a long process of accumulation of good habits where it becomes possible to incorporate JIT. If for example you go over to the system overnight, I expect that it is impossible.

In the first two years the Japanese management introduced the followng changes:

- improved housekeeping, emphasizing the four S's – this included the use of before-and-after photographs

- elimination of waste

- management of detail – according to the UK production director the new management focused on detail two orders of magnitude greater than their UK counterparts

- teamwork development

- investment in new plant and machinery

- total preventative maintenance, with extensive training of operators and others to take more responsibility for this

- statistical process control

- training, where operator training was seen as a very high priority

- the introduction of a new style of production manager, promoting people with detailed knowledge of the process to be managers – 'They replaced the old-style managers who . . . saw themselves mainly as managers separate from the technical function'.

- development of flow in production, including moving to a flow-oriented layout, investment in materials handling and moving machines from a functional grouping to be part of the flow

- better quality of components and material

- setting detailed performance objectives and continuous improvement

- moves to organizing as a single manufacturing team rather than a series of profit centres

Further plans include introducing andon lights and boards, developing greater workforce flexibility and mobility and reduction in set-up times to reduce batch sizes.

What this example highlights is the integrating effect of JIT, bringing together functions and building them into a tight and effective team. Beginning with some simple techniques with clearly visible benefits, the process of gradual culture change can develop as the firm moves into more complex applications and techniques. Therefore early investment in preventative maintenance, total quality and set-up time reduction can provide the foundations for more active involvement and improvement along several dimensions. As many commentators are fond of pointing out, just-in-time is a journey, not a destination.

Before we leave this chapter it is worth reiterating that, despite its undoubted power and effect in dealing with many of the manufacturing problems of the 1990s, JIT is not a substitute for advanced manufacturing technology. There is clearly a limit to the extent to which existing machinery can be improved, and the power of computer-based information systems as control and planning tools cannot be ignored. Significantly, the way in which Japanese firms coped with the threat of the high yen was not only to apply the principles of JIT even more rigorously but also to invest heavily in flexible manufacturing technology and computer-aided design.

JIT is not a substitute for AMT but a necessary first step. By challenging the existing arrangements and practices in manufacturing, it enables much of the ground to be cleared ahead of AMT investment, and it begins to set up the more integrated structures and flexible skilling which AMT requires. A common prescription for moving successfully into AMT is to 'simplify, integrate and then automate'. JIT provides an extremely valuable template for the first and second steps.[38] But, beyond that, JIT needs to be married to appropriate use of AMT; a point underlined by Kochar in his study of Japanese firms, where he comments that '. . . once a saturation point has been reached in terms of methodological improvements, advanced technology becomes very important'.[39]

Notes

1   Ingersoll Engineers, *Integrated Manufacturing*, (IFS Publications, Kempston, Bedford, 1985).
2   R. Schonberger, *Japanese Manufacturing Techniques*, (Free Press, New York, 1982).
3   T. Ohno, quoted in *Engineering Computers*, September 1986, p.58.
4   This list derives from that identified by Shingo Toyota and is discussed extensively in K. Suzaki, *The New Manufacturing Challenge* (Macmillan, London, 1988).

5   C. Lorenz, 'The crippling cost of just-in-case', *Financial Times* 10 January 1986. Others give a higher estimate; for example *Industrial Computers*, August 1986, puts it at £41 billion.

6   Cited by T. Maclean, 'JIT demands a new accounting system', *Works Management*, February 1988.

7   R. Kaplinsky, 'Restructuring the capitalist labour process: some lessons from the automobile industry, *Cambridge Journal of Economics* (1989).

8   C. Voss and L. Ozaki-Ward, *The Transfer of Production Management Techniques by Japanese Companies from Japan to the UK*, Warwick Manufacturing Roundtable Working Paper, October 1987.

9   Schonberger, *Japanese Manufacturing Techniques.*

10 A. Dear, *Working TowardsJust-in-time*, (Kogan Page, London, 1988).

11 R. Kaplinsky and K.Hoffman, *Driving force: The Global Restructuring of Technology, Labour and Investment in the Automobile and Components Industry*, (Westview Press, Boulder, Colorado, 1988).

12 Voss and Ozaki-Ward, *Transfer of Production Management.*

13 J. Burbidge, 'A synthesis for success', *Manufacturing Engineer*, November 1989.

14 J. Burbidge, 'JIT for batch production using period batch control', *Proceedings of the 4th European Conference on Automated Manufacturing*, (IFS Publications, Kempston, 1987).

15 Y. Monden, *The Toyota Production System* (Industrial Engineering and Management Press, Norcross, Georgia, 1983).

16 K. Suzaki, *The New Manufacturing Challenge*, (Macmillan, London, 1987).

17 Voss and Ozaki-Ward, *Transfer of Production Management.*

18 S. Shingo, *A Revolution in Manufacturing: the SMED System*, (Productivity Press, Cambridge, Massachusetts., 1983).

19 Shingo, *Revolution in Manufacturing.*

20 Suzaki, *The New Manufacturing Challenge.*

21 Suzaki, *The New Manufacturing Challenge.*

22 The TPM evolution programme, Japan Institute of Plant Maintenance, 1982, cited in Suzaki, *The New Manufacturing Challenge.*

23 R. Schonberger, *World Class Manufacturing*, (Free Press, New York, 1987); and H. Wildemann, 'JIT production in West Germany', *Proceedings of First Just-in-time Conference*, (IFS Publications, Kempston, 1986).

24 K. Fukuda, *Japanese Style Management Transferred: The Experience of East Asia*, (Routledge, London,1988).

25 C. Voss and S. Robinson, 'Application of JIT manufacturing techniques in the UK,' *International Journal of Operations and Production Management*, 7(4) (1987), pp.46–51.

26 Reported in *Engineering Computers*, September 1986.

27 *Engineering Computers*, September 1986.

28 Schonberger, *World Class Manufacturing.*

29 *Financial Times*, 17 September 1986.

30 C. Voss and N. Harrison, 'JIT in the corporate strategy,' *Proceedings of the 4th European Conference on Automated Manufacturing*, (IFS Publications, Kempston, 1987).

31 Cited in *Works Management*, January 1988.

32 R. Lamming, 'For better or worse', in *Managing Advanced Manufacturing Technology*, ed. C. Voss (IFS Publications, Kempston, 1986); and D. Macbeth and G. Ferguson, 'Strategic aspects of supply chain management', in *Proceedings of 5th International Conference of the Operations Management Association*, University of Warwick, Coventry, 1990.

33 T. Maclean, 'JIT demands a new accounting system', *Works Management*, February 1988.

34 M. Wheatley, 'Easier said than done', *Financial Times*, 4 January 1989.

35 Voss and Harrison, 'JIT in the corporate strategy'.

36 W. Daniel, *Workplace Industrial Relations Survey*, (Policy Studies Institute, London, 1986).

37 Voss and Ozaki-Ward, *Transfer of Production Management Techniques*.

38 Voss and Harrison, 'JIT in the corporate strategy'.

39 A. Kochar, *Design and Operation of Manufacturing Systems in the Japanese Industry*, University of Bradford, Department of Mechanical and Manufacturing Systems Engineering, June 1988.

9 Total Quality Management

9.1 INTRODUCTION

For many years the top priority for manufacturers in Europe and the US has consistently been 'producing to high quality'.[1] That it does not feature quite as highly in Japanese priorities reflects mainly on the enormous efforts already made in Japan to deal with this question.

It is not an exaggeration to say that quality is the most important of the non-price factors on which manufacturers are trying to improve. Hill suggests that it has moved from being a desirable 'order-winning' characteristic to being an essential 'order-qualifying' one.[2] In other words, unless a firm is able to deliver high quality it will not even be able to sit at the same table as the rest of the players in a particular market.

The extent of the quality crisis which has been facing Western manufacturers since the 1970s is highlighted repeatedly in surveys and comparative studies. For example, in 1973, 12 per cent of US consumers felt that Japanese cars were of better quality than US ones, but by 1983 that figure had risen to 40 per cent.[3] Similar figures were given for consumer electronic products. In Garvin's detailed study of the air conditioner industry in Japan and the US, he found that US failure rates were between 500 and 1000 times higher than those of their Japanese competitors.[4] In the case of semiconductors, attention to quality improvements in the production process enabled Japanese producers to obtain an average yield of 50 per cent in the manufacture of memory chips (a market which they now dominate) whereas US firms could only manage 15 per cent.[5]

There is little doubt that quality has moved to the forefront in terms of its influence on the decision to buy a product or service. In 1988 an opinion poll carried out on behalf of the UK Department of Industry commented that '. . . Quality is perceived as the single most important factor in evaluating a company, both by the general public and by industry'. Feigenbaum reports that in the late 1970s only 40 per cent of US buyers ranked quality as being at least as important as price in their buying decisions, but that the 1990s figure is over 80 per cent and rising.[6]

235

9.2   WHAT IS QUALITY?

We ought perhaps to begin with a few definitions of 'quality'. The dictionary says it is:'the degree of excellence which a thing possesses'. John Ruskin, the nineteenth-century painter and art critic, makes a valuable additional point: 'Quality is never an accident, it is always a result of intelligent effort'. Pirsig suggests that quality is not a physical attribute, nor a mental concept, but something embodying both:[7] 'even though Quality cannot be defined, you know what it is'.

In general, it involves some perception of excellence, of reliability, and often an association with higher cost – 'you get what you pay for'. It is also important to see quality as a perceived characteristic which varies between individuals. A commonly used definition is that quality is 'fitness for purpose';[8] that is, the extent to which a product possesses characteristics which suit the users' purposes. It also includes some consideration of reliability, as an indicator of the long-term continuation of this state of fitness for purpose.

This user-oriented approach is helpful in focusing attention on the customer rather than the producer, but it can be argued that it needs some modification. In particular, as Garvin[9] points out, it does not deal with two key problems:

- how to aggregate what may be widely varying individual perceptions of quality, to provide something meaningful at the level of the market

- how to identify the key product attributes which connote quality

An alternative set of definitions emerges from considering the producer's side, a set concerned with establishing standards and measuring against them.[10] *Quality of design* represents the intentional quality which the designer wishes to see produced in order to meet his interpretation of the customer's needs. It is a multi-attribute definition, but has the advantages of permitting measurement against each of these attributes to assess whether or not the intentional quality level has been achieved.

Associated with this is *quality of conformance,* which represents the degree to which the product, when made, conforms to the original design specifications. The extent to which this can be achieved will depend in turn on the various elements of manufacturing; people, processes, equipment, incoming raw materials quality, and so on. This equates to Crosby's idea of quality as 'conformance to requirements'.[11]

The quality process can be seen in figure 9.1. Essentially, the market needs are translated into product strategy which in turn feeds through to the R & D function to design a suitable product and associated specification. It is against this that quality can be measured in terms of conformance to that specification. On the process side quality will be affected by two things; the overall capability of the process (to hold tolerances, and so on), and the way in which quality is controlled within the process. The degree to which conformance to specification can be achieved will depend on these two factors.

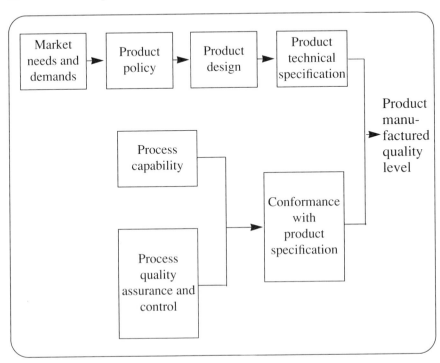

Figure 9.1   Factors determining conformance to specification

Garvin[12] identifies five categories of quality definitions, ranging from the 'transcendental' (such as that offered by Pirsig, which see quality as an innate property of 'excellence' which is also timeless and beyond short-term fashions), through product attribute definitions and the above user and producer views to definitions which try to put a value or a cost on quality or its absence. This last group of views attempt to link quality to the price that people are prepared to pay for it, and the expectations which follow.

But quality is not just about satisfying the customer in the marketplace – it also has an impact inside the firm. With material costs often representing 40–50 per cent of total manufacturing costs any scrap made will have a major impact. Scrap and rework arise from a variety of quality related factors; poor and defective incoming supplies, problems built in at one stage in the process which are not detected until later, problems in handling and transport within the works which lead to damage, and so on. And anything which does 'escape' the notice of the quality inspectors and is passed on to the customer may result in warranty claims, complaints and other public relations problems.

Taguchi[13] takes this approach in his definition of quality, which he relates to the loss to society as a whole as a consequence of not paying attention to quality improvement. His 'loss function' approach takes into account definitions relevant to both producers and users, and helps to provide some kind of cost–benefit accounting for quality improvement activities.

A final point about quality. It is hard to put limits on the term, but it refers to much more than simply the physical attributes of a product. The phrase 'a quality organization' implies a great deal more about the approach taken to dealing with all sorts of other issues – customer care, product design, interpersonal relations, financial management, and so on. So improving quality management may not simply reflect a change in the way product manufacture is treated: it may result in development of the whole organization.

9.3   COSTS OF QUALITY

Calculating the cost of quality is an interesting exercise. At first sight many firms would assume it equated with the costs of running a quality control department plus the scrap and rework costs. But, as Crosby[14] has notably pointed out, those costs are just the tip of the iceberg. The true cost of quality is the total cost due to factors such as:

- disruption to production
- time and resources spent correcting mistakes
- materials, energy and resources wasted on producing the original mistake
- investments in specialist functions to track and find quality problems
- warranty claims
- poor customer relations and consequent costs of advertising and so on to put that right

In other words, the true cost is not just the cost of putting things right. Overall quality costs are made up of two types; those incurred in prevention of mistakes and in checking for them, and those incurred through a failure of these procedures (either in the factory where the result is scrap or rework, or in the marketplace, where they lead to complaints, warranty costs, and so on). Figures vary but it is generally accepted that the costs of quality can be between 20 per cent and 40 per cent in manufacturing, and between 35 per cent and 50 per cent in services.

Some examples lend weight to this, and to the benefits which can follow from 'getting it right first time' instead of having to fix the quality problem. For example, in 1984 IBM estimated that $2 billion of its $5 billion profit was due to improved quality – not having to fix errors. In a 1988 survey carried out by the UK computer manufacturer, ICL, 89 per cent of series 39 users said they would recommend their system to others. In the previous year (before ICL initiated a major quality programme) this was only 67 per cent.[15]

9.4   A BRIEF HISTORY OF QUALITY

Quality in the West

In the earliest days of manufacturing, quality was essentially built into the work of the craftsman. For example, the notion of 'taking a pride in work' was a central pillar of the medieval guild system whereby concern for quality was trained into hearts and minds from apprenticeship onwards. But the Industrial Revolution destroyed much of this one-to-one identification with the product, and led to a loss of the craft ethic and its gradual replacement by the factory system. Although quality was important, especially in the pioneering applications of new technologies evident in the bridges, machinery and other products of that period, it was often in competition with the demands of high productivity to satisfy massively expanding demand.

During this process the location of responsibility for quality shifted. Instead of lying in the hands and judgement of the individual craftsman it began to be built into manufacturing processes and equipment. The aim was to design and mechanize the variability out of the process. It was also during this period that specialization began to emerge as a consequence of the factory system of organization, and with it the idea of separating out responsibility for quality from the worker actually producing the product.

In the latter part of the nineteenth century the focus of attention in manufacturing shifted to the US, where Taylor's ideas were of particular importance. Characteristic of US factories of the late nineteenth century was machinery which had become increasingly sophisticated and capable, but which was not matched by suitable skills. The US lacked the traditional pool of skilled labour which could be found in Europe, and so new models of organizing production began to emerge instead. In essence they involved extensive mechanization and automation, the organization of production along the lines suggested by 'scientific management' (with its emphasis on task fragmentation and standardization, trying to design out errors in product and producer) and the emergence of the specialist department with responsibility for quality. In Taylor's model of the effective factory, quality was one of eight key functions identifed as of critical importance for shop foremen to manage,[16] while Radford's influential book, *The Control of Quality in Manufacturing*, published in 1922, placed further emphasis on the task of inspection as a separate function.[17]

This model became the blueprint not only for the mass production facilities of the 1920s and 1930s but also for many other factories. Typically, emphasis was placed on inspection as the main control mechanism for quality, supporting a process of gradual refinement in product and process design which aimed to eliminate variation and error. It was not a static model, and there were a number of developments in the ideas and techniques available to support quality management. In particular, in 1931 Walter Shewhart wrote a book, entitled *The Economic Control of Manufactured Products*, based on his experience in the Bell Telephone Laboratories. This study of methods for monitoring and measuring quality marked the emergence of the concept of *statistical quality control* (SQC), as a sophisticated replacement for the simple inspection procedures of the 1920s.[18]

This was timely, since the requirements imposed by the Second World War meant that stringent quality control became essential. Emphasis was placed not only on productivity but also on quality control, and from the lessons learnt in applying SQC and other approaches several notable ideas emerged. Of particular interest was the work of a group of quality experts including William Edwards Deming and Joseph Juran (who worked for a while with Bell Laboratories in the Quality Assurance Department set up by Shewart): the group were involved in wartime training and helped to establish the American Society for Quality Control. Within this forum many of the key ideas underpinning quality management today were first articulated, but their impact was limited and little understanding of quality control principles extended beyond the immediate vicinity of the shop floor.

In 1951 Juran published his *Quality Control Handbook*, in which he highlighted not only the principles of quality control but also the potential economic benefits of a more thorough approach to preventing defects and managing quality on a company-wide basis.[19] He suggested that failure costs were often avoidable, and the economic pay-off from preventative measures to reduce or eliminate failures could be between $500 and $1000 per operator – what he referred to as the 'gold in the mine'. A few years later, Armand Feigenbaum extended these ideas into the concept of 'total quality control', in which he drew attention to the fact that quality was not determined simply in manufacturing but began in the design of the product and extended throughout the entire factory.[20] As he put it, '. . . the first principle to recognise is that quality is everybody's job'.

These ideas were not taken up with any enthusiasm in the West, but they did find a ready audience in Japan which was facing the twin problems of catching up with Western practice and rebuilding its shattered industrial base. Much of the reason for the relative lack of interest amongst Western firms can be traced back to economic factors. For most firms, the 1950s were a boom period – the era of 'you've never had it so good'. One consequence of this relatively easy market environment was that the stringencies of the war years were relaxed and there was a general slowdown in effort, in both productivity growth and quality improvement practices.

In the 1960s the concept of 'quality assurance' began to be promoted by the defence industry, in reponse to pressure from the NATO defence ministries for some guarantees of quality and reliability. This grew out of work on 'reliability engineering' in the US which led to a number of military specifications establishing the requirements for reliability programmes in manufacturing organizations. (Some indication of the size of the problem can be gauged from the fact that in 1950 only 30 per cent of the US Navy's electronics devices were working properly at any given time.)[21] Such approaches were based on extensive application of statistical techniques to problems such as that of predicting the reliability and performance of equipment over time.

Quality assurance is the name given to the set of systems (embodying rules and procedures) which are used by a firm to assure the manufacture of quality products. Although a good idea in principle, by 1969 it had become enshrined in an increasingly bureaucratic set of rules and procedures which suppliers needed to go through to obtain certification by defence agencies. Consequently, in the firms themselves quality assurance became in an increasingly dogmatic, bureaucratic and specialized function – a book of rules rather than a live principle.

There were some positive results from this move, notably the emergence of the concept of supplier quality assurance (SQA). In order to ensure compliance with increasingly rigorous standards, certification and checking of suppliers began to take place, in which the onus was placed upon suppliers to provide evidence of their ability to maintain quality in products and processes. Such vendor appraisal was often tied to the award of important contracts, and possession of certification could also be used as a marketing tool to secure new business because it provided an indication of the status of a quality supplier.

By the mid- to late-1970s there were many of these SQA schemes in operation, all complex and often different for each major customer. As a result suppliers faced a major task in trying to ensure compliance and certification. Such congestion led to the need for some form of central register of approved schemes and some common agreement on the rules of good QA practice. There are now a number of national and international standards which relate to the whole area of quality assurance, and require the establishment and codification of complete quality assurance systems. For example, in the UK British Standard BS5750 (equivalent to the international ISO 9001 (1987)) provides a specification for such a system. It requires the setting up of continuous improvement mechanisms, and the maintenance of standards and procedures which are periodically checked. BS5750 emerged in 1979 and was strongly promoted as part of the National Quality Campaign in 1983; despite this emphasis it is estimated that there are still less than 10 000 firms with full certification.

There is no doubt that such procedural approaches have made a contribution to improving quality levels in the West. However, they still represent a traditional view which sees quality as the province of specialists and as something which is primarily controlled through inspection at all stages – as indicated in table 9.1.

Area	Controlled by
Suppliers	Supplier quality assurance
Goods inward	Goods inwards inspections
Production	Stage inspection and tavelling inspection
Testing	Technique evaluation and final inspection
Packing	
Despatch	

Table 9.1   The traditional approach to quality control

During the 1960s, and particularly the 1970s, it became clear that Japanese firms had not only managed to shake off their image of offering poor quality products but had actually managed to obtain significant competitive advantage through their improved performance in this field. This led to a renewed focus of interest and effort in the quality area and the beginnings of adoption of Japanese practices. Garvin[22] reports that the Martin Corporation managed to supply a defect-free Pershing missile one month ahead of schedule in 1961, a remarkable achievement at a time when extensive inspection and testing was the norm, and defects were accepted as almost inevitable by final customers. Of particular significance was the fact that this had been achieved by focusing all employees on the common goal of 'zero defects'. As the company management reflected: 'the reason behind the lack of perfection was simply that perfection had not been expected'.[23]

This led them, and others, to experiment with ways of building worker involvement in programmes which were designed to promote higher quality consciousness and the desire to do things 'right first time'. The first Western quality circle was established in Lockheed in 1975 and others quickly followed. But firms soon began to realize that there was no instant plug-in means of providing better quality – and many early QCs failed after early success. Firms gradually recognized the need for more of a company-wide approach which included operator involvement and a total system approach to quality management.

New tools helped this process; in particular, the idea of statistical process control, which had been developed in the 1940s but which became easier to implement in total systems which stressed operator involvement. SPC, which was applied extensively in the early 1980s, not only improves the control of quality but, importantly, changes the location of responsibility. It brings control of the quality back to the point of manufacture rather than leaving it at the end of the chain. Other approaches include the idea of operator involvement, of top management commitment, of quality as a concept being applied to much more than just the product, the extension of problem-solving techniques beyond the quality area – in short, to company-wide quality control or *total* quality management.

Quality in the East

In the East the pattern of development has been somewhat different. Early industrialization in Japan was not linked to scientific management but to a more much older culture which owed much to the feudal nature of pre-industrial Japanese society. Notions of obligation and loyalty were effectively coupled with the Samurai/Zen tradition of excellence which can be seen in Japanese art and craft.

Despite the devastation of the Second World War, this model of industrialization was not destroyed with the physical elimination of Japanese industry but, instead, adapted and absorbed new ideas and models from the West. In particular, concern with quality was given a high priority, and in the seven years of US occupation experts such as Deming, Juran and Feigenbaum found that their ideas of SQC and TQC were enthusiastically taken up. The earliest inputs of US ideas on quality came from courses offered by a small group of engineers (many of whom came from Shewhart's Bell Laboratories) within a group known as the

Civil Communications Section, in the immediate post-war period. Significantly, as Garvin points out, emphasis in these courses was laid on quality as a key factor; for example, the course manual made the point that '. . . the primary objective of the company is to put the quality of the product ahead of any other consideration. A profit or loss not-withstanding, the emphasis will always be on quality.'[24]

In 1948 the Japanese Union of Scientists and Engineers (itself only formed two years earlier) formed a quality control research group, and invited Deming to give a series of seminars. These were extremely influential, espcially in introducing some of the statistical approaches, but also in encouraging a systematic approach to problem-solving. So successful was his visit that the Deming Prize for quality was initiated in 1951 in his honour. Pride in quality became a key norm in the post-war development of Japanese industry, and state support was also present in the form of the Industrial Standardization Law of 1949, which was derived from the attempts of the Ministry for International Trade and Industry (MITI) to improve the range of products being made and sold.

The early 1950s saw the growing trend towards SQC being applied across the organization, backed up by formal procedures and standard-ization. It is important to note that this trend was led by engineers and middle managers, and was not necessarily seen as a key strategic development by senior management at the time. The concept of company-wide quality control really emerged during the late 1950s, as new mechanisms were developed and as the tools of statistical quality control were applied systematically. Once again, ideas which had origi-nally developed in the West were influential here. Joseph Juran visited Japan in 1954 and laid considerable emphasis on the responsibility which management had for quality planning and other organizational issues concerned with quality, while Armand Feigenbaum came two years later with his message about company-wide quality control.

One lesson that emerged from this experience was the need to involve those in the production process much more, to teach them *why* as well as what they had to do to guarantee quality. A key feature of this is the idea that operators are much more than simply interchangeable resources, as they are represented in the Taylor/Ford model. As Kaoru Ishikawa, son of one of the founders of the Japanese quality movement, said:[25]

> . . . if Japanese workers . . . were obliged to work under the Taylor system, without encouragement of voluntary will and creative initiative, they would lose much of their interest in work itself . . . and do their work in a perfunctory manner.

In many ways this is an obvious point. After all, the likely consequences of treating people as 'cogs in a machine' include:

- disinterested operators

- increased defects in products

- drop in labour efficiency

- no quality consciousness (why bother?)

- increased absenteeism

- increased labour turnover

Total quality management seeks instead to build on worker involvement, creating mechanisms for increasing their ownership and responsibility for quality and providing the training, tools and the top management commitment to support it. In particular, the evolution of the concept included components such as quality audits, quality control circles, problem-solving groups, promotion agencies and the interaction with other alternative approaches to management and organization such as JIT. The idea is – again – to reduce or eliminate unnecessary waste, whether in inventory accumulation or poor quality, and the underlying philosophy is one of continuous improvement. This is achieved by systematically searching out and revealing problems, followed by equally systematic solution using the creativity and experience of all involved.

This Japanese approach is exemplified in the seven basic principles of total quality control:

- get it right first time (control the process while the product is being made)

- make quality easy to see (use displays and indicators to highlight progress)

- insistence on compliance (quality comes first, output second)

- line stop (give responsibility and authority to any operator to stop production to correct quality)

- correct your own mistakes (quality is *everyone's* problem and you own a part of it)

- 100 per cent check (sampling lets bad parts escape)

- continuous improvement (project by project)

Some indication of the power of the Japanese approach can be seen in the fact that Japanese firms aim for 'zero defects' and can actually talk about parts defective per million, while in the West the norm is still 'percentage defects'. It is important to set this in a cultural and political context. Quality has been a prime target in Japan for 40 years, and the overall consciousness of its importance is highly developed. Whereas in the UK, for example, it has been necessary for government to try to push the idea of quality (through ventures such as the National Quality Campaign, and through financial and advisory support for quality improvement programmes), in Japan the drive has always come from the private sector and from the professional associations such as the Japanese Union of Scientists and Engineers. There are some 80 Quality Control conferences every year and total attendance is over 60 000 – and all of these run on a self-financing basis.

Although late in the day, Western firms are now trying to reach the same point as the Japanese in awareness and use of quality technology and techniques. Ironically, there is little about the Japanese system – at least, in terms of tools and techniques – which is peculiarly Japanese. Indeed, much of it – the ideas of SQC and company-wide quality control, for example – derives from the teachings of Western experts such as Deming and Juran. Juran's own comments on attempts to place the credit for the high quality achievements of the Japanese on his (and Deming's) work are helpful here:[26]

> I am agreeably flattered but I regard the conclusion as ludicrous. I did indeed lecture in Japan as reported, and I did bring something new to them – a structured approach to quality. I also did the same thing for a great many other countries, yet none of these attained the results achieved by the Japanese. So who performed the miracle?

9.5   QUALITY MANAGEMENT: TOOLS AND TECHNIQUES

What are the options now open? Effective quality management depends on the systematic application of a handful of basic principles such as statistical process control. However, within the firm itself, the ways in which these techniques can be applied will vary enormously. Innovations in the quality area also combine hardware, software and organizational change, as a glance at the key elements below will indicate.

Price[27] suggests that there are four key questions surrounding quality in organizations, and each of these leads to a particular set of tools:

Question	Tools
1 Can we make it OK?	Process capability analysis
2 Are we making it OK?	Statistical process control
3 Have we made it OK?	Acceptance sampling
4 Could we make it better?	Quality improvement

Statistical quality control

Much of the initial work on quality has concentrated on the questions 2 and 3 above. These can be answered in most cases not by checking everything but by the careful and systematic use of various statistical techniques. SQC is a simple concept which has been extensively refined during the past 50 years, and it is an area in which information technology can make an important contribution through its data capture, storage and processing capabilities. The essence is shown in figure 9.2, where it is broken down into two areas, acceptance sampling and process control.

Each of these is broken into two parts, attribute sampling (which examines some attribute of the product such as its visual attractiveness or its ability to function correctly) and sampling based on measurements of variables (length, temperature, and so on).

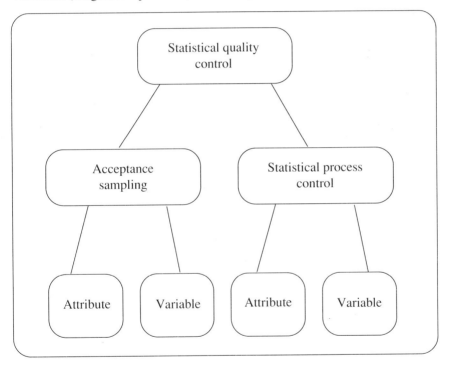

Figure 9.2   Statistical quality control

Acceptance sampling

Acceptance sampling is the basis of traditional inspection-based QC approaches and is essentially retrospective. Various elaborate procedures for making the acceptance decision, choosing the sample, and so on, have evolved, and these techniques are still important where there is no guarantee of incoming quality from previous stages or from suppliers further back up the chain. But the main problem is that it deals with quality problems after they have been built in. Acceptance sampling uses either 100 per cent inspection in which problems are found straight away, and the probability of being correct is 1; or a sample may be taken which is less than 100 per cent but is still sufficient to ensure that only an 'acceptable' number of defects get through. This percentage is set using an operating characteristics curve, which depends upon statistical techniques to calculate the probability of error with different-sized samples.

Statistical process control

By contrast, process control, attempts to monitor and improve things while they are still happening. It looks for deviations from the specification and tries to work out where they have come from. There are two basic sources of variation in any process; random variations (which will always occur) and those which are due to some controllable feature of the operation (whether in the operator, machine, material, and so on). The objective is to monitor continuously (via inspection, sampling, measurement, and so on) and there are many different tools and techniques for carrying this out, some of which are susceptible to automation, others to a manual involvement.

In essence, SPC is a set of procedures for testing the hypothesis that 'the process is still under control and any variations are due only to random causes'. If this is not proven, then corrective action needs to be taken. The basic way of checking this out is via a control chart. This is a simple visual record which allows regular measurements to be plotted. Careful statistical analysis in preparing the chart means that it is possible to identify ranges which take into account random variations, so that when a reading – or, more importantly, a series of readings – moves into a warning or action zone then something needs to be done.

Most common is a 'measured variable' chart, which is prepared during a specially controlled period (to ensure as far as possible that all variations are random and not due to some defect of process or material). Statistical analysis of several samples is used to plot the mean and then to set up warning bands to either side, and then beyond that action lines. The statistics used are such that the chances of a sample being outside the warning bands and in the action area are so low that there must be something wrong, and action must need to be taken. Such a chart is illustrated in figure 9.3. Once this chart has been prepared, samples are plotted on it to test whether or not the process is under control.

An alternative is an attribute sampling chart. Here there is only a choice of 'good' or 'bad', so the statistics are simpler. In essence the

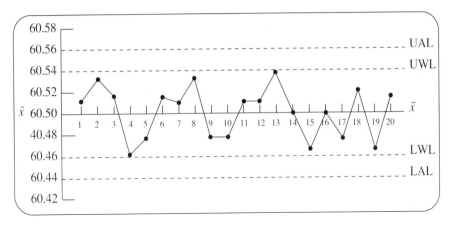

Figure 9.3   Example of a control chart

probability of finding a certain number of defects in a batch can be calculated from theory. This gives us the information to set up a control chart which has limits for warning and action; where if the action area is reached it is because the probability of this happening randomly has been exceeded and something is wrong.

Process capability analysis

It is important to recognize that there are finite limits to the quality which can achieved in any process without changing the process itself. A machine which can only work to a particular tolerance can never produce better than that tolerance – instead it needs to be replaced. All processes are also subject to random variations. and it is only when performance strays beyond these that some form of control action needs to be taken to improve quality. Therefore an understanding of what the limits on process capability are will be essential background information for any programmes based on process control or acceptance sampling.

The generation of such 'baseline' information is the subject of process capability studies. These are essentially variations on the theme of running the machine or process, for a statistically significant period, and then measuring the performance of selected variables or the incidence of good and bad attributes. The relevant control chart data can be calculated from this, so that those responsible for quality control can make meaningful judgements.

Continuous quality improvement

The use of the above approaches can maintain quality at a predetermined level but, in addition, regular analysis of the most common problems or types of defects will give information with which to improve the process or product in some way. A wide range of problem identi-fication and solving tools can be used in this connection, including:

- pareto analysis, which recognizes that 80 per cent of failures are due to 20 per cent of problems, and so tries to find those 20 per cent and solve them first

- histograms, which help to represent this information in visual form

- cause and effect diagrams (fishbone charts), which try to identify the effect and work backwards, through symptoms to the root cause of the problem

- stratification – identifying different levels of problems and symp-toms using statistical techniques applied to each layer

- check sheets – continuously updated check-lists of likely causes, which can be worked through systematically

- scatter diagrams, which plot variables against each other and help identify where there is a correlation

- control charts, which use SPC information to start the analytical process off, asking *why* these errors have occured at this time

Once this kind of understanding of the problem and its contributing elements has been built up, other problem-solving techniques can be introduced. Since the search is on for as many different ways as possible of dealing with the problem, techniques which support creative problem-solving are especially helpful here. Brainstorming and other, related techniques are often used.[28]

Such problem identification and solving is not confined to current problems in quality. Techniques also exist for forecasting possible future problems and for exploring 'what if?' scenarios. A variation of this approach as a formal technique within SQC is Failure Mode Effect Analysis (FMEA), which is used to explore systematically what can go wrong in product and process design, and to find preventative ways of dealing with it, or of minimizing its effect should failure take place.

One key feature of successful problem-solving is the active involve-ment of people who work with the quality problems at all levels. Much of the Japanese success in quality management is based upon ways of ensuring workforce involvement and participation, and the resulting set of skills, experience and individual creativity provides a powerful engine for effective problem-solving.

The notion of continuous change and development is built into quality improvement. This can be characterized as a three-stage process, using the tools of SQC and employee involvement, as shown in figure 9.4.

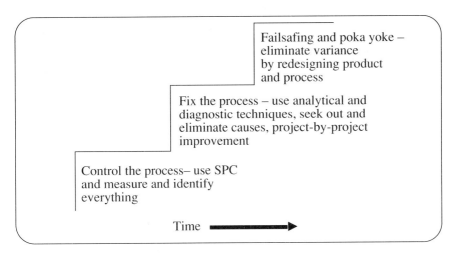

Figure 9.4   Steps towards quality improvement

The concept of poka yoke was introduced in the previous chapter, on just-in-time methods. Its application derives from work carried out by Shingo[29] in Toyota and other companies, and is based on the simple assumption that, no matter how good their intentions, human beings are fallible and will make mistakes from time to time. Consequently, attention should be paid to 'foolproofing' the system in ways which make it impossible for mistakes to occur. As with its application in just-in-time, poka yoke devices need to be simple; for example, a check-list represents a highly effective device.

One last tool which again is attributed to Shingo is the idea of source inspection. Rather than inspection at a late stage in production via acceptance sampling techniques, he argues for systems which try to spot defects at source, at the point at which an error is being made. The principle of everyone taking responsibility for identifying errors and correcting them before they result in defects is critical to this and – once again – depends upon full participation and involvement in the quality issue.

Company-wide quality control

At the core of Japanese quality success has been the idea of *company-wide quality control*, (CWQC) a theme originally articulated by Feigenbaum in the mid-1950s.[30] The basis of this concept is to be able to design, produce and sell goods which satisfy the customer's requirements, and this takes us back to our initial defintions of quality. But CWQC recognizes that there are many dimensions to this, such as:

● customer service

● quality of management

- quality of company

- quality of labour

- quality of materials, techniques, equipment, and so on.

It provides a mechanism, or rather a philosophy, for integrating these elements. In particular, CWQC recognizes that quality is not confined to production and design areas, but that it also involves sales, personnel, accounts and others. The concept is operationalized by generating ideas and methods for solving problems in all areas – and the result is not only improved product quality but improved overall business performance. It uses simple models within which all employees are encouraged to search for all possible means of improving quality. For example, in their training programme the Japan Standards Association ask employees to set up a matrix, plotting the 'four Ms', materials, men, machines and methods, against three undesirables, excess, unevenness, and wastefulness. The resulting analysis includes a check-list which serves as a starting point for directing quality improvements.

The basis of CWQC combines the best of the techniques of SQC (especially in the SPC area) with behavioural techniques for involvement and problem-solving. The key tool is the Quality Control Circle, through which the various powerful techniques of SQC can be applied. CWQC programmes use SPC not in a passive form but in conjunction with active operator involvement, so that the process of watching for quality problems becomes one of continuous improvement, rather than retrospective controlling to a fixed level of quality. The way in which this is achieved is via training, backed up with the regular reinforcement and involvement of the Quality Control Circle.

The process of continuous improvement is also aided by regular quality audits carried out at different levels, from the individual area or department up to company-wide level, and also extended to suppliers.

Quality circles

Quality circles are based on ideas that originated in Japan in the 1960s, building on the work of Ishikawa.[31] They were not applied in the West until a decade later, the Lockheed Company in the US being the first to introduce a circle in 1974 (Rolls–Royce introduced the first circle in the UK in 1978). The concept extended ideas which had been applied in early work on company-wide quality control, especially in the areas of employee involvement and continuous improvement.

The core elements of a QC are simple. It involves a small group (5–10 people) who gather regularly in the firm's time to examine and discuss solutions to quality problems. They are usually drawn from the same area of the factory and participate voluntarily in the circle. The circle is usually chaired by a foreman or deputy, and uses SQC methods and problem-solving aids as the basis of its problem-solving activity. An important feature, often neglected in considering QCs, is that there is an element of personal development involved, not just through formal training but also through having the opportunity to exercise individual creativity in contributing to improvements in the area in which participants work.

The basic activity cycle of a QC goes from selection of a problem through analysis, solution generation, presentation to management and finally implementation by management (see figure 9.5). Once the

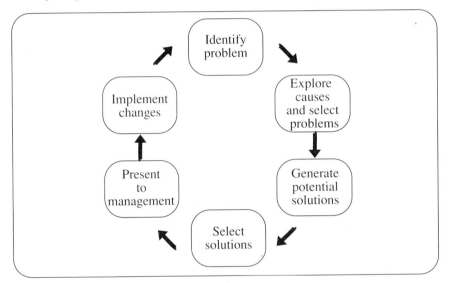

Figure 9.5   The quality cycle

problem is analysed and the root cause identified, ways of dealing with it can be identified. Amongst valuable techniques here are brainstorming (in its many variants) and goal orientation. However, it is important that the structure and operation of the group supports suggestions from anyone (irrespective of level in the organization, func-tional or craft skills background, and so on) and allows for high levels of creativity – even if some of the ideas appear to be wildly impractical at the time. The principles of brainstorming, especially regarding expert facilitation and enforcement of a 'no criticism' rule during idea genera-tion sessions, are important.

The circle does not have to confine itself only to current problems – it can also involve itself in forecasting. Here the possible future problems resulting from each stage can be anticipated and explored. Finally, the group presents the solution to management who are expected to implement it. A key success factor in QC's survival and effectiveness is the willingness of management to be seen to be committed to the principles of CWQC and to act on suggestions for improvement.

Quality circles have been in operation for nearly 30 years and many variants exist. There is, for example, considerable overlap between these groups and the continuous improvement groups mentioned in the preceding chapter. A number of factors associated with success have been identified, including the need to ensure clear roles and to demonstrate top management commitment; for example, by allowing meetings to be held during company time. But there is no fixed model for establishing a circle, nor a perfect recipe for success. Juran suggests that a number of myths have grown up around the concept; these are summarized in table 9.2.[32]

Myth	Fact
QCs only used to solve problems in product quality	QCs can be used to solve problems in safety, productivity, costs, and so on
The QC concept is applicable anywhere	QCs require a significant amount of worker participation in decision-making – managers and supervisors must accept this change
Training is only needed by workers – managers and supervisors already have all the training they need	Training of managers and supervisors must precede that of workers – training must be in the 'why' as well as the 'how' of making QCs work
Adoption of the QC concept requires adoption of the Japanese approach – for example, following their example on time for training	The key need is to match the QC changes to the culture – this will vary from firm to firm
Workers should be rewarded	Rewards, if any, should be matched to the culture of the firm
QCs can be installed purely as a management initiative	QCs must be by voluntary participation– everyone, managers and workers, must believe in the concept and take part of their own free will
The QC can make a major contribution to the company's quality problems	QCs can make a significant but not major contribution – most of the problems are with the system and are the responsibility of managers, supervisors and professional specialists

Source: Lamming/Juran, 1980

Table 9.2   Myths and facts surrounding quality circles

Studies of the success and failure of quality circles often make the point that, while relatively easy to establish, they are often difficult to sustain, and thus the prospects of creating a continuous improvement climate are diminished. McHugh and Dale[33] suggest that two main factors can be associated with the failure of circles:

- they have been introduced in isolation from a broader company-wide total quality programme

- they have been set up with no attention to team-building or development

Much appears to depend upon commitment and support from senior management to a change in the culture of the company; for example, a survey for the Manpower Services Commission (now the Training Agency) in the UK in 1986 reported that:[34]

> . . . the evidence suggests that circles work best when they are considered to be an intrinsic part of total quality management that is supported and driven from within the upper echelons of the organisation by the Chief Executive, and where management are committed to listening to what their employees have to say and want them to be more involved in the business.

Taguchi methods and quality engineering

Taguchi methods were originally developed in the 1950s by Genichi Taguchi, and the approach is now propounded by his son, Shin. Genichi was four times winner of the Deming award, and his methods have had a major impact in the US. The central idea is that while on-line process control is powerful as a means of monitoring and maintaining quality, it does not help to secure the highest quality at lowest cost. Of a product's manufacturing costs, 75 per cent are determined at the design and planning phase; and many of the quality problem solutions can be dealt with cost-effectively at that stage too. Thus the approach is product-centred rather than process-centred.

His definition of a good quality product is one which performs its intended functions without variability, and causes little loss through the cost of using it. His method uses a mathematical expression – the loss function – to measure this loss. This gives a basis for comparing different design options and methods – for focusing improvement.

The basic idea is quality at source, or quality engineering – a form of off-line control to support on-line methods. The more that can be done off-line to eliminate the possibility of variations occurring, the less the on-line control will be needed. There are three basic stages:

- system design – the product and the manufacturing process

- parameter design – this involves experimenting with different parameters in the product design to achieve optimum performance and maximum immunity from variations

- tolerance design – this involves upgrading the product and the process with better elements (materials, equipment, and so on).

The Taguchi method identifies the key control factors which affect performance, but it does so using powerful statistical techniques which quickly help to identify the most important, so that redesign efforts can concentrate on these. This can be seen through the following example. A tile manufacturer was experiencing serious problems with variations in tile dimensions, as a result of having installed a new kiln.[35] It was clear at an early stage that the major cause of this was temperature fluctuations within the kiln, but replacement of the kiln was not an acceptable solution because of the cost. The alternative was to redesign the product, but the question posed was how best to do this to avoid the problem.

Research and analysis produced a list of 128 potential combinations of the seven key factors which they identified as causing the problem. To try each combination would take forever, but the Taguchi method uses a statistical matrix technique to reduce this to a manageable number of experiments. In this case the limestone content was the most important factor, and by increasing the lime content the dimensional variations were eliminated. A second benefit was that the clay content did not have much effect either way – so could be reduced in the tile formulation – and since clay was the most expensive material, the product cost was reduced.

Quality function deployment

A further development on the above ideas is the concept of quality function deployment. This recognizes the fact that quality is something which can be designed and built into a product or service at the earliest stage in its life. All functions likely to have a relevant input during the development and delivery of the product or service are involved, from the earliest stages, in an attempt to improve quality. Such teams have much in common with those involved with 'simultaneous engineering', described in chapter 7.

IT tools

Much of the above has concentrated on behavioural technquies or statistical approaches to dealing with the problem of quality. However, IT has begun to make a major contribution to improved quality manage-ment, both indirectly through the use of more accurate techniques, and directly, by providing a range of data collection and analysis tools. These range from simple devices for measuring, gauging, and so on – often hand-held and functioning as a combined measurement and analysis device for SPC – right through to sophisticated information networks to ensure rapid flow of quality information around a plant – thus setting up plant-wide SPC.

9.6 EXPERIENCE WITH TQM

The increasing emphasis given by manufacturers to the question of quality has paid off handsomely in many cases. Not only are the direct costs associated with quality – scrap levels, time and resource wastage, and so on – reduced, but there are additional benefits such as better customer service and satisfaction which, while difficult to quantify, have a marked impact on competitiveness in the longer term. The following examples give a flavour of the benefits which can be achieved through programmes aimed at total quality management.

1   During the late 1970s the Xerox Corporation found its market share eroded by Japanese competitors, from 18.5 per cent of the copier market in 1979 to 13 per cent in 1981 and 10 per cent in 1984. Through massive efforts in total quality improvement they have managed to reverse this decline, pulling back to 12.8 per cent in 1987 and 13.8 per cent in 1988.

The corporation began the process of change by looking at the Japanese approach through its subsidiary, Fuji–Xerox, and trying to understand the reasons for the dramatic differences in performance. They began to develop benchmarks on several factors, such as number and quality of suppliers, design, product performance, manufacturing processes, and so on. These established the goals towards which the first stage of the improvement programme would aim, and there then followed a major training and development exercise which raised awareness of quality issues. Implementing a change in operating culture included setting up a JIT programme, passing responsibility for quality to all staff and backing this up with the authority to stop the line. Suppliers were reduced from over 5000 in 1982 to around 400 in 1989 and the reject rates for incoming components fell from 3 per cent to around 300 parts per million. The initial results included an increase in customer satisfaction of 38 per cent, a 50 per cent cut in manufacturing costs (mainly achieved through better design) and an improvement in overall product quality (measured by defects per 100 machines) of 93 per cent. The process is now established and supports further continuous improvement in all aspects of quality.[36]

2   A similar story can be found in the case of Hewlett–Packard, which also operated a Japanese joint venture, Yokogawa Hewlett–Packard. In this case the plant not only performed less well than other

Japanese competitors in the late 1970s but was also below other HP plants in the world. Five years later they had achieved significant improvements and won the prestigious Deming Prize for quality in 1982. Their efforts were particularly concentrated upon quality engineering, and emphasized product design using Taguchi methods. The ideas were so successful that they have subsequently been applied to all parts of HP, yielding a direct saving on reduced warranty costs of around $400 million and a further $400 million in savings due to design for manufacture improvements.

Significantly, although quality improved during the 1983–88 period (defects fell from 900 parts/million to 100), the pattern of improvement began to level off, and more effort is now being expended to establish quality improvement principles throughout the company. One important part of this shift, driven by the desire to cut the estimated 25–30 per cent of quality-related costs associated with less than perfect quality, is the move away from an internal definition of quality to a customer-oriented one based on Juran's 'fitness for purpose' definition. For a company such as Hewlett–Packard, founded on technical excellence and engineering skill, this shift towards a customer-led definition implies a major turnaround in culture.[37]

3   Ishikure[38] reports on the turnaround of an ailing US tyre manufacturer when taken over by the Japanese firm Bridgestone. On initial inspection Bridgestone personnel decided that the low productivity and output were fundamentally related to poor quality performance. Surveys on image research confirmed this perception: the company's tyres were rated at a low level amongst a wide sample of customers, and it was clear that this perception grew out of experience of poor quality and reliability. The result was that targets were set to improve quality such that within four years a four-fold increase in output and sales would be achieved. The slogan supporting this was 'quality today will result in quantity tomorrow'.

First steps were associated with simple improvements such as house-keeping and awareness-raising about causes of quality problems. A major programme was then launched, based on what the company termed the '4M' approach – machinery, methods, manpower and material. Over a two year period improvements or changes were made to 530 items covered under these headings. Amongst other changes was a shift from repair to preventative maintenance, and a move towards source inspection at every stage. However, as Ishikure points out:

Manpower proved to be the most difficult problem. The improvement of machines, materials and methods was not enough to produce high quality tyres. It was also essential to instil the concept of quality in the minds of all employees. To achieve this we had to alter past behaviour patterns.

Considerable efforts were expended in changing the culture of the company away from 'us and them' divisions between different levels in the hierarchy, and towards team-building and shared ownership of the quality problem. This was backed up by extensive training in the principles of quality management and especially the ideas of statistical process control. As with many quality programmes, the cycle of 'Plan–do–check–action' was established as the responsibility of every member of the company; rather than separated out into planning by management, action by workers, checking by supervisors, and so on. Quality circles were established to build upon the growing contributions made by employees to identifying and solving problems in the quality area.

Significantly, a major culture shift was in the appreciation that it is a good thing to report problems. Under the old system, mistakes and errors were rarely reported because of the dominant fear amongst employees of being reprimanded. In the new system, reporting of mistakes was actively encouraged, such that by 1983 around 30 incidents per month were being reported. Each of these provided a valuable clue about where to focus problem-solving attention.

The results of the programme were impressive: by 1985 tyre quality was at a comparable level with Bridgestone tyres produced in Japanese plants, and the US market were rating the product as the best in the field (an accolade which was also achieved for the following three years). Production volume increased three-fold, with a corresponding increase in sales. Productivity doubled and percentage defects were reduced by 50 per cent.

9.7   PARTIAL AND TOTAL QUALITY MANAGEMENT

The above cases, and the more general experience of Japan, suggest strongly that significant benefits can be obtained in the area of quality improve-ment. However, as with our examples in earlier chapters, it is clear that implementation of quality management innovations is often concentrated at the substitution end of our innovation spectrum rather than at the integrated end. That is, it involves the use of new techniques which enable better, more accurate control of variables, but it excludes more fundamental changes to organizational structure or process. Or, again, it concentrates on one visible aspect such as quality circles without changing the rest of the culture to support quality as a company-wide value.

Piecemeal innovation of this kind may explain why, for example, there is such a high failure rate in the area of quality circles, where relatively few last beyond six months. The preoccupation with techniques rather than a total philosophy is also commented on by Illidge et al., who attribute some of the blame to the traditional emphasis on quantitative and directly measurable techniques.[39] The importance of changing the company culture to underpin other approaches to managing quality was highlighted in a recent Gallup poll in the US on behalf of the American Society for Quality Control. Of the executives who responded, 87 per cent identifed corporate culture as the most important element in achieving high quality, followed by employee motivation (83 per cent) and employee education (72 per cent) – the formal techniques of quality management, such as SPC, came a long way behind at 53 per cent.[40]

Despite this growing perception of the importance of *managing* quality rather than simply throwing tools at the problem, the gap between Western manufacturers and Japan is still wide. A survey in the UK, for example, found that only 56 per cent of companies have a senior executive who is directly responsible for quality.[41] By contrast, companies such as Toyota have been operating high-level quality teams since the mid-1960s.

9.8   IMPLEMENTING TQM

Putting a total quality management programme into place is clearly not a short-term activity, but requires extensive commitment as a long-term strategic goal. In order to appreciate the key points of such a programme it will be useful to look briefly at the advice offered by some of the 'gurus' of the quality field, each of whom has developed a powerful training approach to the implementation of total quality management.

William Edwards Deming[42]

Deming was one of the pioneering US experts who took the quality message to Japan in the 1950s. He first lectured at the Japan Union of Scientists and worked hard to convince senior Japanese industrialists. His contribution to changing the image of Japanese products from cheap and low grade to high quality was marked by the award to him in 1960 of the Emperor's Second Order Medal of the Sacred Treasure. The citation said that the Japanese people attributed the rebirth of their industry to Deming's work. His work is also recognized in the Deming Award for outstanding achievement in quality improvement, a prestigious annual prize with similar status to the 'Oscar' in the West, which is awarded (amidst extensive media coverage) to the company which achieves the best quality performance.

Deming's basic philosophy remains the same as in the 1950s, arguing that much of the problem lies in management's hands and the need to adopt a totally new approach. He takes the view that more than 85 per cent of problems are caused by imperfect systems or processes, not by workers. The responsibility is clear: workers work *in* systems, whereas the manager's job is to work *on* those systems, improving and redsigning them to function better. His approach to implementation is summarized in his 14 key points:

1 *Consistency of purpose* – setting the mission as the constant improvement of products. processes and services, allocating resources for long-term development in these areas even at the expense of short-term profitability.

2 *Adopting a new philosophy,* which involves a culture change to one in which 'acceptable' levels of defects are unacceptable, and where continual improvement is essential.

3 *Stop mass inspection* – eliminate the need to check by building quality into the products and processes, and carry this process back up the supply chain. Depend on process control.

4 *Change supplier policy* – end lowest-bidder contracts and move to awarding contracts on quality and other non-price factors. Reduce number of suppliers and work with them – certificate on quality.

5 *Continuous improvement* – identify problems due to the system (85 per cent) and to the operators (local faults). Identify and eliminate waste.

6 *Establish training programmes* and put emphasis on education (especially on-the-job training) for all.

7 *Improve supervision* – provide higher quality supervision, not as a control but as a facilitator, helping people do their job. Train supervisors in techniques of quality management.

8 *Drive out fear* – encourage effective teamwork and two-way communication. Make the quality culture a non-punitive one in which it is a good thing to report problems or to stop production until they are solved.

9 *Break down barriers between departments* – establish company-wide ownership of the quality problem. Encourage sharing of information and teamwork; for example, design, production and marketing getting together to improve a product.

10 *Get rid of slogans* – if most of the problems are due to the system, then telling the workers to improve things without also giving them the tools and the understanding of how to do this will be unhelpful.

11 *Eliminate productivity standards and targets,* which often act against quality. Get away from arbitrary levels and encourage statistical techniques to replace work standards.

12 *Encourage pride in workmanship* – restore the craftsman ethic. Reward good work.

13 *Encourage education* – develop the organization by developing the people in it.

14 *Ensure top management commitment*, to establish the mission and to maintain momentum. Quality is a journey, not a destination.

It can quickly be seen that this list is as much a prescription for radical culture change as it is a framework within which particular quality improvement tools can be disposed.

Joseph Juran[43]

Joseph Juran is another recognized architect of Japanese quality improvement. Also taking ideas from the US, he too received the Order of the Sacred Treasure for his work. Company-wide quality management is the concept most often associated with him, although he too brought powerful statistical techniques from the US. He sees the roots of the Western problem as lying in the mechanistic and hierarchical approaches of Taylor and Ford, whose ideas of separation of planning from execution boosted productivity but took away ownership of the quality problem from the worker.

Juran's basic philosophy is to see quality as a key business function which should form part of the strategic plan. Each year goals should be set and resources allocated to achieving them – and performance against these should be reviewed. Adopting a project-by-project approach develops the internal skills and experience to do this in a self-reinforcing fashion, building on success.

This process requires several things; top management commitment, clear communication and ownership of the quality goal and company-wide training about why as well as how to improve quality. The idea of the 'quality trilogy' is valuable here – he sees this as an analogy of the finacial management system in a business.

- *Phase one is planning* – comparable to budgeting. Here the emphasis is on identifying customers (internal and external) and their needs, establishing goals which meet these needs, developing products and processes which can serve these needs, and proving that the process has the capability to meet the goals.

- *Phase two is control* – analogous to cost/expense control. Here, the emphasis is on choosing what to control and how to control it (measurements, standards, and so on) – and then doing it and feeding back the results so as to do it better.

- *Phase three is improvement* − cost reduction/profit improvement in the financial analogy. Here the project-by-project approach comes in, stressing a cycle of diagnosis, a search for root causes rather than symptoms, developing remedies, implementing them and review. In the longer term this also involves making sure that they stick, and that progress ratchets upwards.

Armand Feigenbaum[44]

Like Juran and Deming, Feigenbaum went to Japan in the 1950s when he was head of quality at the General Electric company. His influence on the early development of quality in firms such as Toshiba and Hitachi was particularly strong, and reached a wider audience through translations of his books and articles setting out the concepts of 'total quality control'. Together with Juran, he is credited with extending the understanding of quality control beyond the application of techniques within a particular function, and with encouraging the active involvement of all functions in a systematic approach to quality control and improvement.

His approach is often translated into a function-by-function analysis of what has to be done to ensure quality as a product moves from initial design through development and manufacturing and finally to despatch. It recognizes that at each of the critical stages several functions have an input to make, and that all share the responsibility for quality. As with Juran and Deming, Feigenbaum places considerable emphasis on the role of senior management in maintaining and supporting such a system.

Philip Crosby[45]

Crosby's work is more recent and has had a major impact on Western firms; perhaps because, unlike the 'Japanese trio' described above, his activity has mainly been in Western companies. His major 1979 book, *Quality is Free*, is based on his work within ITT as internal quality consultant. He has become particularly associated with the so-called 'zero defects' philosophy based on the 'four absolutes' − the four key principles. According to these, quality is about:

- *conformance to requirements* − an objective rather than subjective definition, based not on some vague notion of goodness or excellence, but rather on answering the question 'Does it do what it is supposed to do?'
- *prevention, not appraisal* − fix the problem before it happens
- *zero defects, nothing less* − challenging the notion of 'acceptable' levels of failure (after all, as he points out, if midwives were to achieve a 1 per cent incidence of dropping babies on the floor this would still be unacceptable − yet we accept the idea of percentage defects all the time in industry)

- *measurement of the total cost of non-conformance* (not an index, or a formula, such as parts per million or per cent defects) – it is how much it costs you to put it right, often 20 per cent of manufacturing and 35 per cent or more of service-sector operating costs

Crosby's key facilitating mechanism is the *quality improvement team*, a specially established group which is responsible for monitoring and managing all aspects of quality improvement. In particular, he argues for the concept of 'error cause removal', whereby all workers are encouraged not only to seek the root causes of problems ('What stops me doing this job to the highest quality?') but also to follow this through, even if it means challenging the way things have traditionally been done. Crosby also advocates a 14-step improvement programme, in which the similarities to Deming, Juran and Japanese best practice are clear:

1 *Management commitment* – a formal quality policy and mission statement about high-quality, defect-free products

2 *Establish a quality improvement group*, involving representatives from different departments and at senior level

3 *Measurement* – audit quality in each area by formal objective measures

4 *Identify the true cost of quality* to the organization and its components

5 *Improve awareness of quality* across the organization – establish ownership of the problem

6 *Begin corrective action* – set up systems (quality circles, and so on) for improving on the quality problems identified in 4 and 5

7 *Zero defects planning* – explore proposals in the quality improvement group to move towards zero defects

8 *Employee education* on why and how to improve quality

9 *Zero-defects day* – a company-wide event which focuses attention on setting the new zero-defects performance standard

10 *Goal setting* – individual and group action planning for specified targets for improvement within specified time periods (for example, 30, 60 and 90 days)

11 *Establish an error cause removal system*, so that people can tell management safely about what they see as stopping them achieving their goals (see 10)

12 *Recognition* – rewards and recognition to assert the positive value in this company of high quality

13 *Set up quality councils* – to bring quality management people together regularly

14 *Do it all again* – repeat the continuous improvement cycle

Once again, the similarity in ideas with those of Deming, Juran and others is clear. Themes such as the creation of a different set of values which encourage error reporting and reduce the fear of punishment associated with that activity, the emphasis on training and on total company-wide involvement, the stress placed on top management support and the long-term continuous improvement message are all critical features.

9.9 SUMMARY

From the above it is clear that above all success in implementing a quality improvement programme depends upon the extent to which it is seen as a *total* activity, running company-wide and involving everyone in ownership of the quality problem – and the responsibility for solving it. Unfortunately, it lends itself to extensive partial innovation of the substitution variety, using more advanced techniques or tools to support local functions only, or using them in an isolated way, rather than as part of a company-wide package of change. In particular, the challenge to changing the culture of the organization is often shirked; not just because of the considerable effort which this involves but also because of the perceived threats to the *status quo*.

In the end, the organization and management of quality requires a new kind of organization, one which communicates and owns the problem in integrated form. This model is similar to the one we saw earlier for supporting advanced and integrated IT applications, involving similar patterns of networking, and of decentralization of responsibility within an integrated framework where people own the problem and share in solving it. Rather than emphasising structures, processes and a culture which is geared to doing the same thing, day after day, the new organization needs to find ways of becoming a 'learning organization' able to adapt and develop in a cycle of continuous improvement.

Notes

1   A. de Meyer and K. Ferdows, *Quality Up, Technology Down*, Working Paper WP 88/65, INSEAD, Fontainebleu, France, 1988.
2   T. Hill, *Manufacturing Strategy* (Macmillan/Open University, London, 1984).
3   Cited in D. Garvin, *Managing Quality* (Free Press, New York, 1988).
4   D Garvin, *Managing Quality*.
5   G. DeYoung, 'Do it right the first time', *Electronic Business*, 16 October 1989.
6   A. Feigenbaum, cited in DeYoung, 'Do it right the first time'.
7   R. Pirsig, *Zen and the Art of Motorcycle Maintenance*, (Bantam, New York, 1974).
8   J. Juran and F. Gryna, *Quality Planning and Analysis*, (McGraw-Hill, New York, 1980).

9   Garvin, *Managing Quality*.

10 Juran and Gryna, *Quality Planning and Analysis*.

11 P. Crosby, *Quality is Free*, (McGraw-Hill, New York, 1979).

12 Garvin, *Managing Quality*.

13 G. Taguchi, 'Introduction to off-line quality control', Central Japanese Quality Control Association, Magaya, Japan, 1979; cited in J. Oakland, *Total Quality Management*, (Pitman, London, 1989)

14 Crosby, *Quality is Free*.

15 Quoted in *Computing*, 20 October 1988.

16 F. Taylor, *The Principles of Scientific Management*, (Harper, New York, 1911).

17 G. Radford, *The Control of Quality in Manufacturing*, (Ronald Press, New York, 1922).

18 W. Shewhart, *Economic Control of Quality of Manufactured Product* (Van Nostrand, New York, 1931).

19 J. Juran, *Quality Control Handbook*, (McGraw-Hill, New York, 1951).

20 A. Feigenbaum, 'Total quality control', *Harvard Business Review*, November 1956, p.94.

21 Cited in Garvin, *Managing Quality*.

22 Garvin, *Managing Quality*.

23 J. Halpin, *Zero Defects* (McGraw-Hill, New York, 1966); cited in Garvin, *Managing Quality*.

24 Garvin, *Managing Quality*.

25 K. Ishikawa, *What is Total Quality Control? The Japanese Way* (Prentice-Hall, Englewood Cliffs, New Jersey, 1985).

26 J. Juran, 'Product quality: a prescription for the West', *Management Review*, July 1981, p.61.

27 F. Price, *Right First Time* (Gower, Aldershot, 1986).

28 For a useful review of creative problem-solving techniques, see T. Rickards, *Creativity at Work*, (Gower, Aldershot, 1988).

29 S. Shingo, *Zero Quality Control: Source Inspection and the Poka Yoke System* (Productivity Press, Stamford, Connecticut, 1986).

30 Feigenbaum, 'Total quality control'.

31 Ishikawa, *What is Total Quality Control?*

32 R. Lamming, based on J. Juran 'International significance of the QC circle movement', *Quality Progress* 13, November 1980.

33 J. McHugh and B. Dale, 'Quality circles', in *International Handbook of Production and Operations Management*, ed. R Wild (Cassell, London, 1988).

34 B. Dale and J. Lees, *The Development of Quality Circle Programmes*, (Manpower Services Commission, Sheffield, 1986).

35 *Automation*, May 1989

36 H. DeYoung, 'Back from the brink', *Electronic Business*, 16 October 1989.

37 R. Haavind, 'Hewlett–Packard unravels the mysteries of quality', *Electronic Business*, 16 October 1989.

38 K. Ishikure, 'Achieving Japanese productivity and quality levels at a US plant', *Long Range Planning*, 21, (5) (1988), pp.10–17.

39 R. Illidge et al., *Total Quality Management and Organisation Change*, Working paper, School of Economics and Accounting, Leicester Polytechnic, 1987.

40 Cited in *Industrial Computing*, October 1988, p.20.

41 MORI, cited in *Industrial Computing*, October 1988, p.20.

42 W. E. Deming, 'The roots of quality control in Japan', *Pacific Basin Quarterly*, Spring/Summer 1985.

43 J. Juran, *Juran on Leadership for Quality*, (Free Press, New York, 1989).

44 Feigenbaum, 'Total quality control'.

45 Crosby, *Quality is Free*.

10 Inter-firm Linkages

10.1 INTRODUCTION

Up to now, we have been looking at the changing pattern associated with technology *within* the firm. But a feature of the new pattern now emerging is a radical shift in the nature of relationships *between* organizations. In particular, there is a move away from what might be termed 'power-based' relationships, in which there is some kind of hierarchical dependence, towards more of a *network* model in which there is a sense of mutual development within a partnership.

The distinction can be seen more clearly if we contrast the kind of factory operated by Henry Ford in the 1920s with the emerging picture in the car industry today. In works such as Ford's River Rouge plant, the emphasis was on total integration – within the firm – of all operations, from the mining of iron ore to painting the finished Model T cars black, driving them out and selling them. But gradually that pattern has shifted to one in which there is increasing subcontracting of components, materials and services. For many years this process operated on a limited basis and was driven by price; the major car assemblers held a dominant position by virtue of their purchasing power, and their suppliers depended upon them for survival.

However, recent years have seen this pattern change, both in terms of the extent of the network and the basis on which the relationships are founded. Whereas price was the dominant feature until the mid-1970s, it is now accompanied by a series of non-price factors, with quality perhaps in first position. Pressures to reduce inventory have led to a growing interest in just-in-time delivery; a process which operates well in Japan but which requires a radically different form of relationship between buyer and supplier.

A second, powerful force for change lies in product technology. The increasing complexity of many products means that no single firm can be a specialist in all the technologies required for their production and, instead, the skill required will be to bring together different elements produced by independent specialists into a coherent whole. This also emphasizes a trend towards modular design, whereby complex systems such as motor cars are made up of subsystem blocks supplied by a range of different specialist manufacturers. Once again, there are far-reaching implications for the relationships which operate between these firms. If, for example, the systems integration role is to be played by assemblers in the future, then this requires close co-operative work with major suppliers from the earliest stages of the design process – essentially an integration at the strategic level which binds the firms together in pursuit of a common destiny. At the very least such a model blurs the lines between operations.

A third factor is the danger of loss of competitiveness in an integrated arrangement. As Lamming points out, referring to the car industry, the main

> 'drawback of integrating a components supplier is that the combination of a protected relationship with the parent assembler and reluctance on the part of other assemblers to deal with a competitors subsiduary leads over a period of time to non-competitive status for the components supplier'.[1]

At first sight this appears to represent a return to the highly integrated processes typified by the River Rouge plant in the 1920s, in that all activities within the manufacturing and distribution chain are tightly coupled. The difference this time is that the model is not a single highly integrated firm but a well coupled network of many firms, each playing a key role in supporting the others. Within this model the powerful new opportunities opened up by IT for rapid and widespread communication are a critical feature.

10.2   NEED PULL

Inter-firm relationships are not confined to materials procurement but extend into a variety of other areas including design, financial information exchange, expertise (consultancy), specialist services and distribution and marketing. As we have seen, the changing manu-facturing environment places pressure on firms to offer much higher levels of responsiveness and performance on key non-price dimensions, such as flexibility and quality. These have implications for the inter-firm chain as much as for the firm itself. For example:

1   Being able to deliver high quality, defect-free products to the customer depends upon an attitude of total quality management within the firm. Part of that depends on being able to include the suppliers to the firm in that process, so that incoming materials are defect-free. Where problems occur they are dealt with as shared problems, and there is a continuous improvement cycle operating

at the interface as well as within the firm. Such a total quality supply chain depends on information and openness.

2   A rapid response to customers, offering short lead times or just-in-time, depends upon being able to acquire materials just in time for manufacture – but without the cost penalty of carrying excess stocks. As we saw in chapter 8, this can theoretically be achieved by having responsive and flexible suppliers, but in practice the demand is for a very different kind of customer–supplier relationship in order to make this work.

3   Delivery accuracy depends on being able to ensure that supplier firms are given a firm forecast against which to manufacture. But in reality customer circumstances change in response to an uncertain environment and so there are always fluctuations in what appear to be firm plans. The next best thing to certainty is the ability to communicate rapidly and effectively the requirements and changes, so that the supplier firm can use its internal flexibility to cope.

4   Design and product customization depend on being able to translate customer needs and wishes into suitable form – and to capture any changes rapidly and correctly. For complex projects in which a wide range of skills and experience must be brought to bear, the need is to establish relationships which permit free, two-way exchange of information, and to support these with rapid and accurate communication so that each designer is working with the same information.

What emerges is the need right across the manufacturing chain for closer coupling, approaching a single highly responsive system. Whereas the 'traditional' model involves what can be termed 'arms-length' dealing and minimal contact, the new model requires much more blurred lines between firms. As shown in figure 10.1, the need is also for communication to be established at a variety of levels throughout both organizations – in short, to make the boundaries between them more permeable.

It is relatively easy to argue this theoretically, stressing the powerful force for convergence which the uncertain manufacturing environment represents. For example, the idea behind just-in-time supply is to reduce delivery times and inventories, and to increase responsiveness in the system as a whole by a process in which each supplier becomes flexible to its customers by requiring its suppliers to deliver just in time right back up the chain. But it is not easy to achieve this in practice, and this co-operative chain often breaks down. This happens for a variety of reasons; for example, there may be strong and weak links in the chain, as the relationships between firms are rarely on the basis of equal partnerships but of differing levels of power. Strong customers can put pressure on weak suppliers using, amongst other weapons, the threat of taking their business elsewhere. At the other end of the chain, major suppliers of basic items – such as steelmakers – can put pressure on smaller customers by refusing to adapt to customer needs.

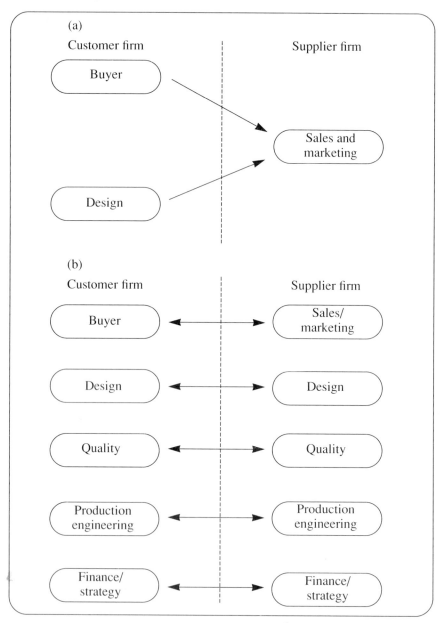

Figure 10.1 Links between firms: (a) old model; (b) emerging model.
Source: **Lamming, 1989**

A similar story can be found in product development. Theoretically, the increasing complexity of many products can best be managed by a high degree of co-operation between specialist suppliers and assemblers, each bringing to bear their particular expertise to produce a high-quality, high performance product and to bring this to market rapidly and efficiently. In practice, such linkages often founder because of reluctance to share information, slow speed and inaccuracy – when such communication does occur – and the underlying climate of mistrust engendered by the kind of purchasing practices described above.

Therefore there is a growing conflict between the need pull on a manufacturing environment, which requires closer working together by all links in the chain to provide what Macbeth calls 'the key deliverables' (price, quality and delivery service), and the reality of existing relationships, which are based on competition and conflict rather than co-operation.

10.3 TECHNOLOGY PUSH

Two technological trends are emerging which can help to meet this challenge. The first is the emergence of IT as a powerful tool to facilitate communication. In the above chains a key requirement in moving to closer relationships is better information flow – getting the right information to the right people at the right time. Traditional paper-based systems are not fast or flexible enough to permit such tight coupling between firms, but IT opens up considerable opportunities.

One example of this is the concept of electronic data interchange (EDI), which is the name now given to the passing of information (intra- and inter-firm) from computer to computer via electronic communication networks. EDI is particularly well suited to routine transactions such as invoicing and purchasing, and has been widely used in areas such as banking, where the settling up of cheques between banks is done electronically, for many years. However, its application in manufacturing is a comparatively recent phenomenon.

In its most basic form, EDI is a way of eliminating paper from internal and external systems. Two features of most routine transactions represent areas in which EDI can contribute significant savings; the elimination of rekeying of the same information into different computer systems, and the reduction of errors. In the case of rekeying, some examples from the car industry illustrate the potential of EDI: General Motors recently calculated a saving of $250 per car, while the UK Rover company in 1987 moved to processing around 15 per cent (140 000 per year) of its invoices via EDI. This was equivalent to some 50 miles of paperwork which was no longer needed![2] In the computer industry, IBM is aiming to cut 7–10 per cent of its administrative costs through the use of EDI, especially in the area of purchasing, where the technology will also assist in the implementation of their version of just-in-time manufacturing. In its network, which has over 100 suppliers connected via EDI, information is exchanged on invoices, purchase orders, shipment notes, engineering graphics, NC data and quality information.

The question of error reduction is also important. Estimates suggest that up to 70 per cent of data is rekeyed into other computers – and each is a source of error. Over 50 per cent of complex documents – for example, a bill of lading – contain at least one error when first presented. EDI can reduce or eliminate this; one example is in the UK Customs and Excise Department, in which an initial error rate of 80 per cent on first submission of paper documents was reduced to less than 5 per cent using EDI.

EDI is, of course, not confined simply to financial or purchasing transactions between firms. Another critical area of development is in design – a feature which we explored in chapter 7. Here, increasing product complexity is leading firms towards ever-closer co-operation and – in theory at least – a powerful aid to this is integrated CAD, in which design information can be exchanged electronically.

But, beyond this, EDI is a way of contributing to overall *effectiveness*, not just local efficiency. For example, it can provide considerable support for the above-mentioned models of closer co-operation by providing a rapid and effective communication system between firms. By linking directly into a customer's production schedule a supplier can forecast more accurately and in more timely fashion for his own material requirements planning and master production scheduling – and can adapt more easily to change if information arrives early enough. Quality, time to market, product performance, customization, and so on can all be supported in similar fashion by closer linkages.

The emerging EDI model of inter-firm contact appears in figure 10.2; it is interesting to note that this implies considerable blurring of the boundary lines between firms.

A brief case study illustrates the way in which EDI can improve inter-firm operations. In the UK the automotive components supplier BTR was invited by Ford to participate in an EDI network. The company, which supplies a wide range of rubber hoses and seals to the car industry, explored various network options, including Ford's own Fordnet and Istel's EDICT, and eventually joined the latter in 1986. Istel charges a £750 implementation fee plus a £950 annual fee. In addition, BTR needed some customization of the software to suit its own data entry and link to its internal computer systems. The system is now typically used to send invoices and other documentation. Benefits include a shorter

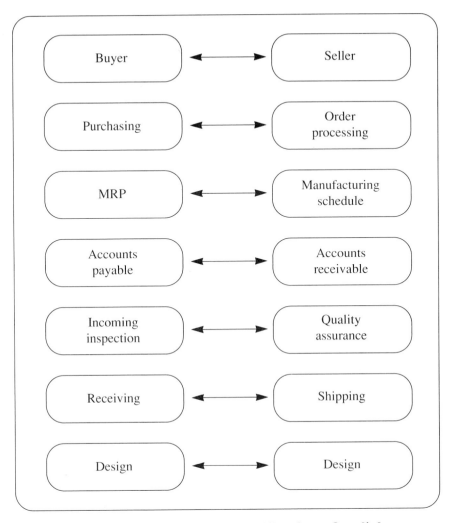

Figure 10.2   The influence of EDI on inter-firm links

payment cycle (and thus better cash flow), and EDI also facilitates smaller stockholding and more responsiveness in the supplier. A further benefit is the reduction of the level of outstanding payments on supplies due to mismatches between invoices and goods received notes (these are down to 10 per cent from an original 25 per cent). However, the greatest benefit to BTR comes not in these invoice-related items, but in the changes EDI makes to the way in which they receive delivery schedules. The system helps them work just in time, because they have the time to plan and react, and communication is more two-way. When changes from the customer side arise, there is sufficient notice to adapt their MRP and other schedules and smooth the pattern for production.

EDI has begun to diffuse rapidly. According to DEC, market growth is predicted to increase dramatically, from an estimated $87 million market in 1987 to an expected $1896 million by 1992. This reflects not only increasing traffic but also more users: 1989 estimates suggest around 5000 users in the US[3] with 1600 in the UK,[4] representing around 70 per cent of the traffic in Europe. Predictions are for increasing growth, especially in engineering; for example, the UK market is predicted to grow at 35 per cent up to a figure of some 5000 firms by 1990.

One of the key factors which will influence diffusion is the extent to which generally accepted EDI standards can be adopted. As with our discussions in earlier chapters, a potential problem with EDI is that a high volume of electronic data travelling around various inter- and intra-company networks will require some form of traffic management system, some set of rules to enable it to operate effectively. The pattern within EDI has been for the gradual emergence within sectors of *de facto* standards; for example, ODETTE for the motor industry, EDIFICE for electronics, CEFIC for chemicals and EDICON for construction. This framework is used by a number of organizations that provide networking and EDI facilities – Value Added Networks (VANs) – which use the principles of industry standards in their systems. A world standard is being developed by the International Standards Organization under the name EDIFACT, but this is not yet fully developed or agreed.

10.4   NETWORKING

The second powerful technological trend is towards networking as an alternative to traditional forms of integration within a manufacturing chain. This is an example of 'technology' used in its widest sense, as a different way of doing things, although it is also an area in which the power of IT as an aid to developing and operating networks is of enormous importance.

Essentially, networking involves an alternative to the traditional models of vertical and horizontal integration, in which a wide range of resources can be focused on a particular area without necessarily involving the centralization of control over those resources. A model based on co-operation rather than control, it has sometimes been described as a value-adding partnership (VAP), defined as 'a set of independent companies that work closely together to manage the flow of goods and services along the entire value-adding chain'.[5]

Its strength lies in the possibility of combining the advantages of small firms – rapid response and high flexibility – with the economies of scale associated with larger firms. Arguably, in an environment requiring increasingly diverse outputs and flexible response this model is better suited than a highly integrated large firm.

The example of the clothing industry

Some examples will help to flesh this out. In recent work which has been carried out on the clothing industry, this shift towards more inter-firm co-operation and networking is apparent.[6] The challenges here are familiar; the need to respond to increasingly demanding fashion markets, the requirement for more frequent product innovation and higher levels of flexibility and customization, and the constant pressure on quality, all underpinned by a major price threat from overseas manufacturers.

Typically, the manufacturing chain in such an industry is long and diverse, including elements such as:

- natural and synthetic fibre suppliers
- spinners
- fabric producers
- dyeing
- printing
- fabric finishing
- garment producers
- accessories
- wholesalers
- retailers

Case studies reveal a number of different permutations of buyer–supplier relationships in this chain, and a pattern which is changing over time. Major retailers appear to split into two distinct groups. The first are those which 'play the market' on a highly competitive, seasonal contract per garment basis. Here continuity of relationship is not important and suppliers may be changed, reselected or deleted according to price, range offered and delivery performance. Such retailers often have no house style, and so coordination of quality and sizing is not important. Most retailers adopt a policy of multiple sourcing, putting no more than 25 per cent of their business with a single supplier, and all relationships are essentially 'arms-length', with little or no concern about how the supplier achieves his targets as long as the right price/quality/delivery mix is achieved.

The problem with such an approach is that it focuses too much attention on price, and this promotes undercutting and competition between rival suppliers to the detriment of overall performance. It is a traditional power-based relationship in which there is an assumption of a 'zero-sum' game, in which there have to be winners and losers.

By contrast, there is another pattern which can be termed the 'interventionist' approach. This is based on long-term contracts, guaranteeing minimum order levels for suppliers which have been vetted and approved. In return for this security, the retailer expects strict conformity to design specifications, quality standards and delivery schedules. In many cases the retailer becomes actively involved in the suppliers processes and may visit and inspect at any time. Information flow to the retailers is also at a high level and such suppliers can be seen as 'locked' into retail chains in return for guaranteed profitability. But they can also be seen as engaging a powerful resource on the retailer side in terms of advice, financial support for new technology, and so on. They also facilitate three-way discussion with powerful suppliers, such as those for fabrics, thereby taking pressure off smaller firms which might otherwise be 'bullied' by the big boys.

These two models represent opposite ends of a spectrum and both have strengths and weaknesses. In many ways the highly fragmented chain has the advantages of speed and flexibility, achieved through largely informal networks and multiple subcontracting. By contrast, the long-term relationship chain tends to be slower moving, but emphasizes high quality and reliable delivery. As to the preferred chain model, much depends on the kind of products and markets involved but there are a couple of common prescriptions for success.

1   Buyers and suppliers should maintain regular formal and informal contact. Formal agreements are a useful point of reference and are likely to succeed for product lines which do not require high degrees of flexibility. However, when used as a form of control over fast track products, they can be cumbersome and encourage partners to cover up the actual situation. In the end there is a need for much higher levels of trust between firms in the chain.

2   There should be points of contact between buyers and suppliers at every point in the chain. For example, one of the greatest weaknesses occurs when the retailer assumes full responsibility for material purchase and selection, breaking the link between manufacturer and fabric supplier.

The case of the Italian textile industry

A model which offers an interesting development, by fusing the advantages of both long-term connection and networking, has been working for several years in the Prato region of Italy.[7] Over the past 20 years, 15 000–20 000 small firms have replaced all but one of the large vertically integrated textile mills, in a clear reversal of the traditional economic logic of scale economy, 'big is best'. By 1982 these firms employed 70 000 people and exported $1.5 billion of products. It is not clear what triggered the process – speculation suggests that it was perhaps to avoid labour legislation – but, whatever its initial cause, it has effectively led to a network model.

The roots of this transition can be seen in the case of Massimo Menichetti, who took over a typical large integrated textile mill from his father in the early 1970s.[8] His company faced typical challenges of the kind we have seen throughout this book; rising internal costs (especially of labour) and a fragmenting and increasingly competitive market demanding greater variety, more rapid product innovation and overall flexibility in response. He decided to try a radical alternative response to create flexibility, by breaking the company up into small independent units and selling up to 50 per cent of the shares to employees in those businesses. The arrangement was that over a three-year transition period the new enterprises would have to make half of their sales to other customers beyond the Menichetti group.

This model began to diffuse more widely in the Prato region, and the pattern is now – at one level – highly fragmented. Whereas in the traditional case an integrated firm would be responsible for every aspect of the business, from market assessment, through design of fabric and products, production and to final marketing and distribution, these tasks are now handled by small units (often family-based) along the chain. Each has developed particular skills in particular areas and work is contracted out to the unit best able to respond.

In this way the network is well placed to react in highly flexible and rapid fashion to a fast-changing market. It does, however, require a co-ordinating role and this is played by an independent *impannatore* – a role analagous to an orchestral conductor. This role essentially involves maintaining the network, facilitating problem-solving and inter-node communication. In particular, the *impannatore* has an overview of the whole picture, being involved with the customer interface, dealing with suppliers and helping units to negotiate on costs and delivery. He also plays a key role in securing the finance necessary for investments in new technology, and in ensuring that the network members receive an equitable share of the profits from their operation.

This model has a number of features which make it philosophically attractive. First, it represents a return to small-scale manufacture along the 'small is beautiful' lines suggested by Schumacher.[9] It allows a much higher level of participation and develops a strong sense of ownership – of both the problems and the rewards of the business. The link between activity and outcome is much clearer, and the people involved have a sense of being in control of their own destiny. It also facilitates the growth of individual entrepreneurship and creativity.

But there are also strong economic reasons to support such an alternative. In the case of Menichetti the individual units had managed to achieve a 90 per cent machine utilization level after five years, with increased labour and machine productivity. Capacity had been increased by 25 per cent and the variety of products had been extended from an average of 600 to 6000 different yarns in the eight units. Inventory levels for work in progress and finshed goods were cut from four months to 15 days. The pattern was repeated in other Prato region networks, such that in the period 1970–82 production of textiles doubled, while the rest of Europe suffered a serious decline.

This experience is not confined to the clothing industry. Other examples include tile-making in Modena and Reggio Emilia, where one interesting feature is the R & D and technical support symbiosis between tile-makers and the small engineering firms who build their machinery.[10] Other sectors such as furniture and shoe-making have also operated such consortium arrangements successfully, and they are diffusing beyond their Italian origins.[11]

This model does not simply involve the repetition of an historical cycle. Certainly, the putting-out system which existed before the days of mass production in the textile and clothing industry was apparently similar, with flexibility of response to fluctuations in demand, fashion, and so on being catered for by the use of a large number of small, centrally coordinated units. But the old system involved a *hierarchy* of control, essentially integrated in a pyramidal structure of dependency relationships. By contrast, the Prato model is essentially *network*-based, utilizing the key property of networks, which is to be simultaneously highly centralized *and* decentralized. Thus each unit can preserve a considerable measure of independence but is also connected into a shared information and resource system which gives it considerable scale economy in the key areas where small size is a disadvantage. The role of the *impannatore* is not that of an owner at the top of a pyramid but of a manager/facilitator. Being at the centre of the network means that a complete overview of the whole operation is available, and the *impannatore* can then act to assist particular parts of the network focusing the total resources of the network on particular problems. The system is knowledge/information-based – not a power-based – on a partnership rather than a hierarchy.

The role of IT in this is important. One reason why such networks can operate effectively is that up-to-the minute information is available across the whole network simultaneously. Knowledge about the state of the market – demands, volumes, fashion changes, and so on – can be effectively matched against available capacity, and the work routed and rescheduled to suit the availability of units. Here again, we see the role of the centre not as a power source but as a coordinator.

There have been a number of criticisms of the model, not least that it represents a unique set of circumstances which could not be applied elsewhere. Yet similar patterns can be found in an increasing number of industries as the shift away from mass production and vertical integration takes place. The car industry, for example, is increasingly becoming a network of specialists as the complexity of the product increases to the point at which one single firm – even of the scale of Ford or GM – no longer has sufficient resources or expertise to produce in integrated form. Instead, components are bought-in from specialist firms, and the car producers concentrate on assembly and marketing. Japanese industry has led in this trend, with estimates suggesting that Toyota now produces only 20 per cent of the value of its cars, with GM and Ford at 70 per cent and Chrysler at 30 per cent.

A case study: Benetton

One of the more famous examples of the networking model, combining new forms of production and distribution organization with the power of IT is the Italian clothing company, Benetton. Originally founded in 1965, it is now one of the largest fashion firms in the world. Benetton chose from the outset to operate a subcontracting network, similar in many ways to the Prato model described above, demonstrating the power of net-working to create a flexible response to a rapidly changing fashion market. The effective size of the firm is very large, giving it economies of scale of a traditional kind – in purchasing, and so on – but the structure and organization of production allows a high degree of decentralization, permitting high-variety, low-volume and rapid-changeover production – essentially offering economies of scope. As Belussi puts it, the firm has created a high degree of mobility whereby it can shift its organizational boundaries, centralizing or decentralizing rapidly to match environmental demands.[12]

Another key feature is the multiple levels of entrepreneurship which this system encourages. At the core is the traditional firm, retaining control over key processes such as design. But on the periphery are various actors – suppliers, subcontractors, sales agents, and so on – all of whom have a stake in developing their own businesses synergistically with the core. This arrangement could be criticized for its dependence on the core for key information and overall strategic direction, but on the 'plus' side, it offers a model in which small independent units can draw upon extensive resources of advice and technology.

10.5 EXPERIENCE WITH ALTERNATIVE MODELS

There is now considerable pressure upon firms within supply chains to move to alternative models which stress co-operation rather than confrontation. In particular, the buyer-driven emphasis on just-in-time (or close to it) in delivery response coupled with greater stress on non-price factors such as quality and design is having a major (if at times enforced) influence on many smaller suppliers. Success in meeting these challenges involves a much higher degree of security and protection in a competitive marketplace. For example, Macbeth suggests that 'if suppliers are competitive in the non-price area, the cost differential needs to be of the order of greater than 15 per cent before alternative sources are seriously considered'.[13]

Despite the lip-service being paid to new concepts and models, progress towards full integration is often partial and slow. For example, Baxter et al.[14] reported on a study of 50 suppliers adopting JIT practices in the engineering and electronics industry. Their main findings were as follows:

1   Most major buyers are actively reducing the size of their supplier base, using a variety of vetting and assessment procedures to achieve this.

2   Quality comes very high on the list of factors of importance, and receives proportionately higher attention in terms of resources. Despite an upward shift in standards expected from buyers, emphasis is not yet being placed on 'total quality' ideas. There is a concentration on catching problems after they have occurred, and far less on proactive systems for prevention and anticipation of problems. This is accompanied by an almost total lack of measurement of the cost of quality.

3   In general, delivery standards here are moving upwards and towards just-in-time response, with fewer parts being delivered more frequently. But this is often a 'brittle' rather than a flexible response; dropping everything to meet hurried demands rather than responding carefully to well thought out customer requirements.

4   Changes are taking place in the customer company: there are a number of examples of new job types or roles in procurement, and a growing emphasis on providing quality assurance support for suppliers, often extending to some form of certification. However, this role is often that of a policeman rather than a builder or facilitator, placing emphasis on day-to-day matters and concerned with 'blame accounting'. There is correspondingly less interest in longer-term strategic perspectives, or the establishment of long-term contracts.

5   Information flows still tend to operate in a one-way direction: 'Suppliers still perceive that parts are put up for competitive tender and the company with the lowest price still wins the order. The customer wants perfect knowledge of the supplier company's business. Their financial setup, the production system, where they source their materials and so on. This is not met with similar candour on the customer's side.'

6   There appear to have been few changes in the internal organization and operation of suppliers to help them become more

competitive in the non-price area. Consequently, any improvement in their performance is not so much due to doing things differently as to running faster and faster on the existing treadmill.

Other examples indicate what can be done to improve practice in relationships between firms. For example, Hewlett–Packard in Bristol, UK buys in 80 per cent of its final product – so supplier links are crucial. The company depends upon high-quality, just-in-time supply and is also a high-volume user, consuming some two million silicon chips per month. It now has a new breed of purchasing manager, whose main concern is no longer with the purchase price of the component but with the hidden costs; for example, of stopped production because of a faulty component. HP have reduced their supplier base down to around 170 firms but are providing a variety of support to these, including long-term contracts and advice; they even pay for essential equipment for suppliers so that they can guarantee quality.[15]

Another firm, ICL, reported at a recent conference that they have only recently begun to try to resolve the matrix of conflicting objectives of price, service and low inventory. This has led them to 'discover' the supply chain as a factor that directly affects their own ability to perform competitively. They now look for factors such as:

- vendor quality and JIT delivery
- short lead times
- quality on time and in time
- design and development on time

10.6   MOVING TO NEW MODELS OF INTER-FIRM LINKAGE

Within a typical chain of manufacturing there might be several players, each supplying to the next. While in theory each depends upon the other and both stand to gain from the other's success, in practice the nature of the relationship is less balanced. There is a constant struggle to gain advantage and control; for example, by a large customer playing off several potential suppliers against each other on prices. In such arrangements information is seen as a power resource, and is rarely shared beyond a minimum level.

The emerging pattern becomes one of inter-firm competition, in which units in the chain see each other in adversarial terms, as part of a zero-sum game in which there will be winners and losers. Weapons in the battle include the delayment of payments, deliveries, quality, and so on. Key factors which determine competitiveness in the overall business – such as quality, speed of delivery and reliability – become threatened by failures in the supply chain, which act to predispose the buyers negatively against the suppliers. Finally, the cycle is reinforced by the apportioning of blame.

This free market trading of goods and services along a chain has been explored by many writers, notably by Williamson, with his theory of transaction costs.[16] The view emerging from this work is that firms will eventually find the costs of managing such an external chain so high that they will bring the activities in house – essentially pursuing a policy of vertical integration. The advantages here include economies of scale and control, but the disadvantages include the fact that managing an internal hierarchy also has associated transaction costs.

A second feature is becoming important in the manufacturing environment of the 1990s. Whereas earlier, more stable environments allowed vertically integrated firms to flourish, exploiting economies of scale, today's increasingly turbulent and fragmented pattern requires firms to focus on distinctive competences. They cannot be all things to all men – not least because of the high cost implications of such specialization across the board. What is needed is more responsive units, each capable of reacting in specialist fashion to environmental changes. The emergence of decentralizing concepts such as 'the factory within the factory' or autonomous production cells or business units represent examples of such a process of change, and their logical extension is into networks of firms. Rather than pressure for more integration, we are actually seeing forces pushing towards *dis*integration.

The value of networks is that they can represent the best of both worlds. A network can be simultaneously tightly coupled so that it behaves as a single large firm – with all the implications for, of scale economies efficient use of resources, bargaining power, and so on – and highly decentralized, with all the implications for flexible response, close contact with customers, manageable scale, innovation, and so on. Organizational cultures within such networks are also important; traditional highly integrated firms often resort to bureaucratic structures and hierarchies, and are dominated by a mechanistic role culture. By contrast, networks permit organic, task-based cultures to operate. They are held together not by power but by shared information and knowledge.

Such models appear attractive on paper, but realizing them in practice is far from easy. Although, on paper, it is possible to demonstrate the rationality of co-operation and sharing along a chain, in practice firms and the individuals within them do not always behave rationally. There is a basic assumption that the game is competitive and that it must have some form of zero-sum outcome with winners and losers. Changing the nature of relationships towards those which emphasize trust and co-operation is not a trivial exercise. Carlisle and Parker[17] suggest a number of essential ingredients:

- a trust that both parties will do what they *say* they will do

- a willingness to risk becoming vulnerable to the other party, supported by a firm belief that the other party will not take unfair advantage

- a sensitivity to each other's needs, and an active dedication to seeing that both parties' needs are met, so far as that relationship can meet them

- a high level of clear and candid communication, which leaves neither party in doubt about the feelings of the other towards the relationship, and the understandings within that relationship

A survey of inter-firm relationships in the UK argued that adaptation on the part of both buyers and suppliers was essential to achieving the improvement in competitiveness now being sought in the marketplace. The requirements being placed on small suppliers should not be seen as hurdles, but as part of a long-term process of continuous improvement in which both buyers and suppliers participate.[18]

This view is echoed by other writers such as Houlihan,[19] who argues that key features of 'co-makership' include:

- mutual confidence with benefits for both parties

- co-operation starting at an early stage of development

- a problem-solving attitude over borderlines

- an integral cost approach

He comments on the lack of a strategic perspective on supply chain management, arguing that the challenges of the new environment will require:

- market reach to customers (such as providing tighter linkage through EDI) and gradually effecting a change from product-oriented to customer-oriented supply chains – in this new model the customer order penetrates deep into the supply chain

- integration back to suppliers – such 'co-makership' requires a changing relationship based on co-operation

- more efficient logistics along the total industry chain

- integration with the internal computer-integrated business through the use of more subtle and effective organization policies

He summarizes the nature of the changes required as follows:

> To support these schemes both customer and supplier have to develop more than the transaction handling systems and networks

that connect their operations on a day-to-day basis. Success requires effective policy, planning and organisation. Without a coherent supply policy, the order is placed without strict economic limits regarding size and frequency. And, without effective co-operative planning between supplier and customer, every order will be an adventure requiring excessive inventories, capacities and administrative support.

Simply to invest in technology is not the answer. No matter how powerful the information technology may be, exploiting its potential for improving communication and control requires major organizational adaptation:

Those companies that invested in order processing systems and networks (costing $8 to $16m to install) are surprised to find that the change to customer driven from product driven supply chains requires major redesign of these systems...the underlying concepts are more at issue than the procedures and technical links.

Amongst the barriers to implementing such new models are:

- management phobia, a blind acceptance of past practice and a dread of the unknown – a typical example of this is the reluctance to move to single sourcing, since that implies too risky a dependence on one supplier

- the legacy of abuse – a history of aggressive cost-cutting and price squeezing leads to a breakdown in trust, and to rebuild these damaged relationships takes time – (and sometimes changes in key personnel)

- parochialism – 'your end of the boat is sinking' – localized protective attitudes and, particularly, internal functional organization which stresses local issues over sharing and co-operative efforts

- the arrogant technocrat who relies on technological solutions, while management abdicates its responsibility and seeks technical 'fixes' – ably assisted by over-zealous suppliers of software and hardware

Lamming[20] provides a detailed analysis of the difficulties of moving from old to new models, for buyer–supplier relationships in the motor vehicle industry (see table 10.1). The industry traditionally operated a model based on features such as:

- closed competition – new business for one supplier can only be won at the expense of another's loss

- one-sided design effort

- restricted information – for example, on the buyer's requirement levels or the supplier's cost structure
- reasonable capacity utilization for both parties
- satisfactory delivery patterns for buyer

Model	Competition	Sourcing decisions	Data/ information	Capacity	Delivery	Price changes	Quality	R & D	Stress level
					Factor				
Traditio- nal up to 1975	Closed but friendly	Price- based bids	Restricted to minimum necessary	Both OK	Large quantity. buyer's choice steady	General negotiation (annual)	*Laissez faire*	One- sided	Low– medium
Stress 1972–85	Closed and deadly	Bids	One-way supplier opens books	Both spasmodic unlinked	Variable quantity. buyer's demand unstable	Conflict in negotiation (spasmodic)	Aggressive campaigns	Shared for cost reduction	High
Resolved 1982	Closed, some collaboration strategic	Quality bids	Two-way: forward build	Gradually improving – linked	Small/ variable quantity. buyer's demand stabilizing	Annual economics plus negotiations	Joint effort	Shared for development	Medium
Japanese ?	Collaboration agreements	Performance history costs	Two-way knowledge of true costs	Coordinated and planned	Small quantity. agreed basis stable	Annual economics plus planned reductions	Joint planning	Shared 'black hole'	Very high

Table 10.1   Lamming's model of changing relationships

Such a 'satisficing' model based on win–lose outcomes survived not least because of two key features, an expanding market and a perception of 'the inevitability of less than perfect quality' – essentially a less than perfect compromise. It persisted until the early 1970s and approximated to what Lamming calls the 'big stick' approach, in which the buyer ruled and dealings were essentially at arms length. It involved a hierarchy based on buyer power which could, if taken to excess, provoke a supplier response to subvert the relationship.

However, this traditional model is almost extinct today since the pressures on competitiveness have forced firms to re-examine links. The next model was the product of the early stages of such an examination. Lamming calls it the 'stress' model and sees it as arising from two sources; technological pressures (created by exploitation of new opportunities) and increased pressures of competition which make survival a key concern. Faced with those pressures, firms looked for reductions in unit costs – and a good short-term way of achieving this was to squeeze suppliers' prices. The balance in the earlier win–lose satisficing model was replaced by a strong win-plus pressure, whereby the buyer not only had to achieve a win on price but also had to try to squeeze a little more out to achieve a saving. Such a policy, coupled with the encouragement of suppliers to compete for the same business, led to 'suicidal' price wars in many sectors.[21]

A second feature of this was the shift in information flow; buyers began to demand the right to look at the supplier's books before agreeing any price increases – but the reciprocal arrangement did not operate.

However, design liasion went against this trend by virtue of the fact that buyers, concern with cost reduction meant that they were willing to listen to ideas from suppliers which might help in this process. But 'the collaboration was wary, however, since neither side trusted the other with its best ideas, for fear of losing them to competitors.'[22]

The period of time during which this model operated did untold damage to relationships, some of which will take a long time to repair.

By the early 1980s, as firms began to move out of recession, a gradual shift towards a 'resolved' model took place. In addition, concepts such as JIT were beginning to diffuse into European industry, and an appreciation of the full implications of total quality policies and the need for supplier support in these was developing. Firms were looking towards more co-operation with suppliers and a shift away from confrontation.

It is one thing to specify the nature of this co-operation, but it is quite another to achieve it, not least because of the damage done to trust during the 'stress' period. Although there was some shift to single sourcing and other forms of co-operation, the relationship was far from well developed, and this is still largely the case today, with a growing need to co-operate and a recognition of the requirement for trust to underpin all of this. The exception appears to be the case of the 'Japanese' model, which developed along a different trajectory but now represents a transferable alternative, offering features such as:

- long-term relationships with single sources

- collaborative research

- linkages – common strategies, equity exchanges, and so on

- 'black hole' design engineering, whereby suppliers' expertise is accepted and the key skill of the assembler becomes one of systems integration – the detailed design of components is often not recorded on the vehicle builder's drawings, the item being indicated simply by a black hole

- a two-way flow of information, supporting pull-based scheduling JIT

Although some of these may appear to be elements of a traditional vertically integrated system (the roots of which can be traced back to Henry Ford) the difference lies not only in ownership but also in the nature of the relationships which underpin it. Whereas old models – even intra-firm models – were often win – lose, the new models are based on win – win, synergistic development, supported by openness of information and mutual trust.

Such models can be applied in the analysis of other sectors. For example, similar patterns, and the difficulties in moving from one to another, have been reported in electronics and clothing.[23]

10.7   MAKING THE CHANGE

To move to such alternative models requires a systematic process of change. A key feature of this is the shift to seeing relationships between firms not as a short-term tactical matter but as a long-term strategic factor. Instead of concentrating on immediate advantages, which focus on improving efficiency by maintaining high levels of uncertainty (through attempts to play off weaker and more dependent firms against each other, by exploiting win–lose strategies and so on), emphasis increasingly needs to be placed on building partnerships which recognize a high degree of mutual interdependence and which open up the possibility of win–win outcomes.

One prescription for achieving this is offered by Macbeth and colleagues.[24] They suggest a six-stage model, led by the major buyers, for implementation of a change away from operational to strategic relationships.

1 Work out a detailed strategy at the corporate level within the buyer firm, which recognizes and agrees on the long-term strategic role of inter-firm relationships. Once developed, this view needs to be disseminated widely within the buyer firm, in order to effect a similar change in attitudes and behaviour across the organization.

2 Implement this internally: part of the problem is often the lack of 'demand pull' for different relationships and this only emerges when the buyer firm has undertaken a close look at its own manufacturing operations. By moving into approaches such as JIT and TQM, the awareness of what is needed from suppliers can be increased.

3 Bring the supplier chain into this framework. Once the pattern is established, the next stage is to involve suppliers in the process and to create with them a process for establishing and maintaining performance standards (via certification) and for facilitating a continuous improvement in the relationship.

4 Create new internal resources to manage this framework. To support the development of new working relationships will require structural changes and dedicated resources to implement and maintain them.

5 Target a pilot scheme on selected items, focusing first on a small range of suppliers and products or services, and using this to iron out problems and increase learning about alternative ways of operating.

6 Widen the pilot scheme to cover the whole range of products and services.

Such a step-by-step programme is similar to the continuous improvement project-by-project approach in JIT and TQM. The benefits which can emerge from such changing relationships are becoming clear, but the difficulties in moving to such models should not be underestimated. The process challenges much of the existing organizational culture 'the way we do things around here' – and this pattern may well have been laid down over tens or even hundreds of years. To change perceptions from win–lose to some form of mutual development, or to move from short-term considerations to longer-term strategic perspectives which may carry short-term costs, is not achieved painlessly. As we will discuss in chapter 14, the requirement for organizations to adapt and change – to become 'learning organizations' – will be critical to achieving competitiveness in the 1990s.

Notes

1   R. Lamming, *The Causes and Effects of Structural Change in the European Automotive Components Industry*, International Motor Vehicle Programme report, MIT, Boston, 1989.
2   Cited in *Computing*, 26 February 1987.
3   Cited in *The Computer Bulletin*, July 1988.
4   Cited in *Automation*, May 1989.
5   R. Johnston and P. Lawrence, 'Value adding partnerships', *Harvard Business Review*, July/August, 1988.
6   H. Rush, M. Whitaker and W. Haywood, '*Its on the Lorry, Guv! Buyer/supplier Relations in the UK Clothing Industry*', Occasional Paper, Centre for Business Research, Brighton Business School, 1989.
7   F. Belussi, *Benneton: Information Technology in Production and Distribution: a Case Study of the Innovative Potential of Traditional Sectors*, SPRU Occasional Paper 25, University Sussex, 1987.
8   R. Jaikumar, Harvard Business School Case 686-135, October 1986.
9   E. F. Schumacher, *Small is Beautiful*, (Abacus, London, 1974).
10 E. Goodman and J. Bamford, *Small Firms and Industrial Districts in Italy*, (Routledge, London, 1988).
11 M. Piore and C. Sabel, *The Second Industrial Divide*, (Basic Books, New York, 1982).
12 Much of the information for this case study is derived from a detailed report by Belussi, – see note 7.
13 D. Macbeth and G Ferguson, 'Strategic issues in supply chain management', in *Proceedings of 5th Annual Conference of the UK Operations Management Association*, University of Warwick, Coventry, 1990.
14 L. Baxter, D. Macbeth, N. Ferguson and G. Neil, 'Management control in supply chain JIT, *Proceedings of 4th International Conference on Just-in-time* (IFS Publications, Kempston, 1988).
15 Cited in *The Engineer*, 24 November 1988.
16 O. Williamson, *Markets and Hierarchies*, (Free Press, New York, 1975). '
17 J. Carlisle and R. Parker, '*Beyond Negotiation: The Customer-Supplier Relationship*, (John Wiley, Chichester, 1989).
18 T Robinson, *Partners in providing the Goods: The changing relationship Between Large Companies and Their Small Suppliers* (3i, London, 1990).
19 J. Houlihan, 'Exploiting the industrial supply chain', *Manufacturing Issues 1987* (Booz Allen and Hamilton, New York, 1987).
20 R. Lamming, 'For better or worse? Buyer–supplier relationships in the UK automotive industry', in *Managing Advanced Manufacturing Technology*, ed. C. Voss (IFS Publications, Kempston, 1986).
21 For example, see J. Bessant, D. Jones, R. Lamming and A. Polland, 'The West Midlands automotive components industry', Report to West Midlands County Council, 1985.
22 Lamming, 'For better or worse?'
23 For examples, see Rush et al. It's *on the Lorry, Guv!*
24 Macbeth and Ferguson, 'Strategic issues'.

11 Total Integrated Manufacturing

11.1 PUTTING IT ALL TOGETHER

In previous chapters we have seen the influence of the changing manufacturing environment, with its increasing emphasis on non-price factors, on the types of technology in use. The problem of delivering these features for firms across the spectrum of manufacturing is essentially one of resolving the productivity paradox – how to be flexible while also making efficient use of factors of production. In order to meet this challenge, firms are searching for equipment and methods which improve quality, increase flexibility and responsiveness and enhance product design and customer service – as well as reducing or at least maintaining costs and thus prices.

A significant weapon in the strategic armoury on which firms can draw to equip themselves for the battle of the 1990s is the new generation of IT-based systems. These offer advantages not only as substitution innovations (improving efficiency in individual operations, and doing what was always done a little better) but also in combination, where they can contribute to improvements in overall manufacturing effectiveness. These trends towards systems integration and convergence were descibed in chapter 2, and in subsequent chapters we have examined some of the main contributing streams. In this chapter we will explore the coming together of these in emerging models of 'computer-integrated manufacturing'.

Our discussion has also focused on 'technologies' in the broader sense of the word, such as just-in-time, total quality management and networking, which offer advantages through reorganization and better use of both existing and new resources. Central to the effective use of any of these technologies is a large measure of organizational change in parallel with technical change – a theme which will be taken up in chapter 12. So the emerging model should be seen not as simply the extension of technical integration but rather as the convergence of three streams; new IT-based equipment and systems, new techniques such as JIT and networking, and organizational change. Rather than simply computer-integrated manufacturing, we should perhaps be talking about total *integrated manufacturing*.

11.2   THE BLUEPRINT

Like many of the jargon words of the new manufacturing environment, CIM is something of a 'Humpty-Dumpty' word; that is, it means whatever its advocates want it to mean. Boden and Dale[1] review the concept under the title 'What is computer-integrated manufacturing?' and identify at least ten distinct categories into which such definitions fit, ranging from narrow extrapolations around a particular technological stream to those which take a total organizational perspective and which include a strategic component. For our purposes it will be useful to refer to our earlier definition of CIM as:

> 'the integration of computer-based monitoring and control of all aspects of the manufacturing process, drawing on a common database and communicating via some form of computer network.

At its heart it involves the convergence of activities at different levels and in different parts or functions in the organization and, through the use of networks, it offers the opportunity for organizational activities to be simultaneously both highly centralized *and* highly decentralized. One of the most common models for CIM is some form of integration hierarchy, as indicated in figure 11.1.

Here the basic-level operations – machine controllers, data collectors, and so on – operate autonomously but also communicate information to level 2, which covers the overall monitoring and control of a cell. This would, for example, be the case with a simple flexible manufacturing cell. Further up the chain, a plant controller would handle the activities of several cells, coordinating their use of resources and monitoring their overall performance. Level 4 would involve the integration of other key functional areas; for example, design and marketing and would represent a shared information system of the kind represented by MRP2. Level 5 would be an overall business systems integration, in which the financial and sales information would be linked into the manufacturing system and level 6 would be the overall board-level strategic view which includes long- and short-term perspectives, and so on.

The common characteristics of this kind of architecture are that they involve *networks* that information can be shared throughout the system. Changes anywhere in the system will update the rest of the information in the system; so, in one sense, the entire operation can be seen to behave as if it were a single enormously complex machine. But this is not simply a centralizing and concentrating process. The key property of networks which CIM shares is the ability to be simultaneously highly centralized *and* highly decentralized. Thus the economies of shared resources and information can be added to those of local autonomy and flexibility in uncertain environments.

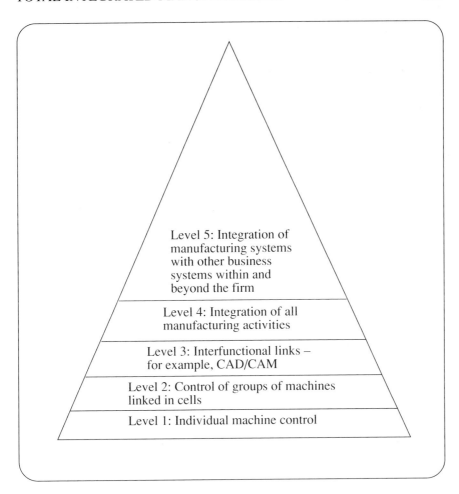

Figure 11.1   Hierarchy for computer-integrated manufacturing

At the level of technological potential, the move to CIM is relatively straightforward, although short-term problems of software interfacing, and so on still persist at levels above 1 and 2. Integration into cells, whether on the shop floor or in other areas such as the design function, has already taken place, and there is increasing integration at the inter-sphere level, to which Kaplinsky refers.[2] This would correspond to a level 3/4 integration, which can be bought from and supported by many AMT suppliers. Beyond that, the integration problem becomes more complex in technical terms, particularly in the software field, but the real challenge lies in the degree to which organizations can adapt their practices, structures and underlying culture to support integrated working. As we saw in chapter 6, for example, having the technical capability to share information between functions is not the same thing as actually exploiting that opportunity.

Two features must be stressed here. CIM is not an off-the-shelf product – it is a concept. What gets integrated with what – and to what level – is very much a local decision related to the specific contingencies of that situation. Successful implementation of CIM depends on understanding this and having the necessary strategy to relate it to the needs of the business – not to fashion, or to keeping up with the competing Jones's.

The second point is that, as this book has hopefully shown, there is more 'technology' available than computer systems. Having identified the business needs and an overall manufacturing strategy, successful firms select their technology from a wide portfolio that includes both technical systems and organisational changes.

11.3 DIFFUSION

We have already looked at the diffusion of individual technologies in previous chapters. In general terms, over 70 per cent of most manu-facturers in industrialized countries have made some form of investment in AMT at level 1 in our model above. But the available evidence suggests that as we climb up the levels, so the level of application falls off rapidly, even in large firms.[3] For example, in a recent (1988) study of the UK the consultants A. T. Kearney estimated that 79 per cent of respondents had at least one CIM component in place, but few had anything approaching full CIM. Even the most comprehensive installation in the UK had only 65 per cent integration, and was using 94 per cent of its functions. Within lower-level systems – for example, a level 3/4 application such as MRP2 – the typical extent of integration for successful users is often around 70 per cent while for many firms it is much lower. In other words, the potential for CIM is still to be exploited.

Of course, not every firm wants or needs to move to the very high levels of integration suggested by the highest levels in our model. Nor do they necessarily need to effect whatever integration they seek through IT. The key issue for the 1990s is to have a strategy for integration up to an appropriate level to support the business in a highly turbulent and demanding environment.

11.4 COMMON FEATURES

Analysis of a variety of cases in different sectors and using different components of CIM suggests that a number of common features are beginning to emerge. Kaplinsky[4] suggests that moves towards 'post-Fordist, restructuring will involve four major systemic changes, each of which has far-reaching implications for management and organizational practice. These are:

- the move from economies of scale to those based upon flexibility and which emphasize demand pull rather than production push

- the changing basis of competitiveness, away from price and towards a range of non-price factors which stress innovation and other product features

- the changing relationships between the parts and the whole – whereas in Fordism the emphasis is placed upon the individual worker, machine or firm, in post-Fordism it is on the systemic interlinking of different elements, whether within the firm (as in a JIT system) or between firms

- the changing role of labour, in which the old view of it as a cost, a necessary evil, is replaced by a view of labour as a key resource, providing flexibility and creativity within new production systems

We can see the emergence of these factors in case studies of CIM implementation. First, the drivers for innovation are increasingly non-price factors viewed at a strategic level, rather than the more traditional short-term efficiency-improving approach which characterizes cost-saving innovation. The use of CIM and its components such as FMS and CAD/CAM represent moves away from dedicated factories and equipment geared, as the old River Rouge works was, to producing one thing in high volume. But the true power of CIM emerges as it exploits the communication and control possibilities in networking to create an integrated system, offering advantages above and beyond those represented by the individual elements.

Integration is further facilitated by the use of production management philosophies such as just-in-time and total quality, both of which stress the need for a systemic change. The culture of the firm is changed through these innovations such that problems which were always someone else's become shared, and the traditional barriers between functions break down. Inter-firm relationships undergo similar restructuring, not just at the structural or procedural level but in their basic nature, moving from confrontation to co-operation.

In practice, it is the coming together of new communication and control technologies with new organizational forms and practices which facilitates early successes in moving towards a post-Fordist model.

11.5   BARRIERS TO CIM

While early evidence suggests that significant benefits can be obtained from CIM, this needs to be set in context. Although there is considerable promise, a number of key issues need to be resolved before this can be realized. In essence these are barriers which have been mentioned in the context of the individual technologies already discussed, but their significance grows as levels of integration increase.

Software and networking standards:

Many of the difficulties in CIM are in the software for linking elements in a network and in the overall integrating philosophies to be used. Physical integration of equipment – such as interfacing robots and handling devices to machine tools – is relatively straightforward; but software integration, the problem of getting different items of equipment to talk intelligibly to each other, is a major difficulty.

What is needed is some form of 'highway code' for all the electronic traffic in an integrated automation system, to clearly establish the various rules of the road – who has priority, speed limits, parking areas, and so on. This would be difficult enough if all the potential inputs of electronic traffic were part of a system made by the same manufacturer, but in practice the requirement must be for an 'open' system that will permit any item of equipment from any supplier to be linked into a factory network of any size. Such an 'open systems interconnection' (OSI) standard is crucial to successful software integration. Once established, producers of automation equipment can design their products accordingly, suppliers of components to them can build special translator chips, and so on – and standardization becomes possible over the whole market.

The fear is, of course, that a single supplier will be able to impose his particular standard and 'lock in' users to his equipment; thus there is concern and argument about which standard to adopt. Specifications for OSI have been agreed at the International Standards Organisation in Geneva, and a proposed seven-layer model has been available in outline form since 1978. This will allow the kinds of interconnection needed in a factory, from basic automation right up to high levels of integration at plant level and beyond.

Having a specification for OSI is not sufficient, however, since that was established primarily for general communications between items of IT. The requirement is for a manufacturing-specific version, and the front-runner here is the General Motors Manufacturing Automation Protocol (MAP), mentioned above. MAP is not a GM product, but a standard specification which suppliers have largely begun to adopt. Most major automation manufacturers, including Fanuc of Japan, have announced that they will support it; and semiconductor firms such as Intel have also begun to produce chips to enable them to implement it. Its likelihood of becoming a *de facto* standard appears high as a result of this, and also because of backing from other agencies.

However, many critics of MAP argue that its adoption for all but the largest system is like using a sledgehammer to crack a nut – it is too sophisticated and powerful. For most cells some form of proprietary local area network (such as Ethernet) will be sufficient; if, indeed, such a high level of integration of communication and control is needed at all. The emerging strategy for many firms is one of 'islands of automation' based on cells, but with care being taken to build 'hooks' into the cells so that they can be joined together at a later stage.

Significantly, evidence suggests that levels of computer integration in Japan are much lower than in Western countries. For example, the US control systems manufacturer, Allen Bradley, estimates that in 1988 the US spent \$20 billion on control software and equipment, Europe \$10 billion and Japan only \$2.5 billion.[5] Part of the explanation for this discrepancy is that emphasis has traditionally been placed on the use of simple systems (such as kanban) and on the use of people as a direct link between different automated elements. Although there is now a major, MITI-led \$1 billion research programme in Japan, designed to develop standards in factory automation, the emphasis is still upon human rather than electronic communication.[6]

Financial appraisal and accounting

One of the major difficulties in considering a *strategic* investment is that it is essentially for the longer term, whereas most investment appraisal systems deal with much shorter time periods. For example, CIM-type systems will require long time periods for payback, often five years or more, whereas the average expectation under normal circumstances in Western industry is two to three years. The result is that investments are either justified on the basis of figures force-fitted to traditional formulae, or not justified at all but entered into as an 'act of faith'. For example, Parkinson and Avlonitis cite one FMS company manager as saying 'if management looks for figures, I can make them up. However, if management in industry in general is committed to financial justification, then the FMS concept is surely on the way out.'[7] This has placed considerable emphasis on the need to develop new techniques, a need highlighted by many writers including Kaplan and Johnson in the US,[8] and Sizer and Motteram[9] and Primrose[10] in the UK.

The argument is about two aspects of accounting, how to justify investments and how to account for them in day-to-day operation. In the former case, there is a need to find ways of building in 'intangible' strategic items into the appraisal: for example, the cost of not being flexible in response in the marketplace in five years' time. Emphasis is shifting, following the Japanese example, to consideration of non-financial criteria where 'factors such as desired market share, cycle time, reject rates and innovation are given more weight in managerial decison-making than calculative exercises about financial viability'. Kaplan cites the case of Yamazaki in Japan, one of the most successful machine tool-builders in the world, which has pioneered the use of FMS. Its Mino Kama plant cost $18 million in the early 1980s and produced extremely good results in terms of lead time reductions, space savings and inventory savings. However, despite these, its return on investment over the early years of operation was only 10 per cent – and thus it would have been rejected by most US or European firms as a poor investment.

Kaplan makes the point that ' ...most of the capital expenditure requests I have seen measure new investment against a *status quo* alternative of making no new investment – an alternative that usually assumes a continuation of current market share, selling price and cost. Experience shows, however, that the *status quo* rarely lasts. Business as usual does not continue undisturbed.' The need is for a new approach from the financial accounting community; but, as in so many cases, there is considerable inertia and commitment to established procedures. He sums up the problem succinctly when he suggests that 'conservative accountants who assign zero values to many intangible benefits prefer being precisely wrong to being vaguely right!'[11]

For example, Primrose argues that alternative bases of justification can be used which allow for a shift away from traditional savings to those of better quality, delivery and shorter lead times. They have developed a computer-based system to help firms in their investment appraisal, but this has not yet diffused widely and most UK firms still rely on traditional (and often inappropriate) techniques for investment appraisal.[12]

Another organization, CAM-I, commissioned a major study on accounting and AMT and found that in investment appraisal, most firms still considered intangibles and qualitative benefits only marginally after looking at direct labour and materials savings. Firms in the sample also tended to look for payback in less than three years. [13] A study of 106 UK companies, carried out by the Centre for Business Research in 1988, found – more optimistically – that 87 per cent attempted some form of inclusion of intangible benefits.[14] However, the sophistication of techniques for doing this was very limited; 55 per cent of firms used only one method of appraisal, and only a handful of firms used more than three methods. Clearly, there is a greater chance of useable estimates if more than one method is used to quantify intangibles.

Increasingly, there is a trend towards involvement of more than one department in the appraisal process, reflecting the broader impact which AMT can be expected to have. In the survey 42 per cent used some form of group discussion to enable them to generate data on intangible benefits, while a further 35 per cent used a detailed study rather than simple cost estimation technqiues.

Some firms have begun to develop alternatives. For example, in the US, Cummins Engines installed a new flexible machining line for a brake product in 1985. At that time the company was looking for new products and markets to offset a maturing of the diesel engine business. They placed a premium on flexibility such that although the cost of a flexible line was two to five times higher than a similar dedicated plant would have been, the tooling would be two to three times cheaper to change. In the event the return on the investment was over 40 per cent, largely because of the inherent flexibility to switch to other products.

In the area of production cost accounting one of the problems is the obsolescence of accounting ratios. Traditional indicators such as standard hours produced, tonnes produced, machine utilization or labour efficiency are now no longer relevant; for example, because of the decline in direct labour or the emergence of just-in-time principles which would argue against using machines unless they are producing something which is actually needed. Instead of traditional costing based on allocating charges to particular machines or operations, some form of flow accounting based on throughput or total factory performance is being preferred.

Supply-side problems

Another major problem facing the early users of CIM has been the immaturity of the supply side. Whereas most firms have experience in selling particular elements of manufacturing systems – computers, software, machine tools, robots, and so on – there is still no turnkey capability whereby one firm can offer the complete range of products and services in a standard package. Instead, each CIM installation is a highly specific application, involving combinations of different elements.

Meeting this need has posed a considerable challenge to the supply side. In the early stages of CIM implementation a number of problems were due to suppliers overselling their capability, especially in the area of systems integration. This led to the emergence of a new service, that of systems integration consultancy, which offered a managing agent role, putting systems together and guaranteeing that they would work together.

From the review of problem issues in previous sections it is clear that users contemplating the implementation of advanced and integrated manufacturing systems are likely to need varying degrees of support in the process. For a few large firms with experience in automation and/or complex products the main resources to support the development of even complex systems can be found in house. The principal external resources needed are equipment (and even here the buyer firm is in a strong and well informed position to specify exactly what is wanted and to evaluate different suppliers against this), and in specialist expertise in narrow problem-specific areas.

At the other end of the scale an inexperienced small-to-medium-sized user firm will need a wide range of support, including:

- technical support in business/technical audit, planning, simulation, and so on to arrive at a suitable configuration – this might include consideration of alternative solutions to the problems facing the firm, such as the introduction of quality programmes or just-in-time

- help with financial and strategic planning

- feasibility study and investment justification support

- education and awareness-raising, especially at board level

- support, perhaps via an external managing agent or project manager, for the implementation, planning and execution of a large-scale project – this latter job would include bringing in the different systems and suppliers, ensuring that they deliver on time, coping with emerging problems, and so on

- training for operation;

- support for organizational development and the management and execution of necessary change

The majority of firms fall somewhere between these two categories. Even for large and experienced users, some form of partnership involving joint problem-solving is to be preferred and this will be essential for the smaller firm. From the supply side, components of this would include:

- aiming to provide solutions to problems, rather than technology packaged and sold as a panacea for all ills

- total rather than partial solutions which reflect supplier bias towards selling a particular product

- help in choosing different approaches to solving the problem – for example, involving managerial or organizational as well as technological change

- incremental approaches which permit gradual change which the organization can afford and, more importantly, can absorb

- guarantees of support and technical service in the long term

There are a number of different actors in the automation market, but so far none are really in a position to supply this range of services and equipment. As one major supplier put it, in their own advertisement: [15]

> . . . the main obstacle to integrated manufacturing is the inherent incompatibility between existing systems. Integration was not a consideration when many of these systems were acquired. There is no simple solution to the problem. CIM is a business strategy, not a single product you can buy. No single supplier has all the products to integrate a manufacturing enterprise.

Although suppliers generally recognize their inability to offer turnkey solutions there are still serious weaknesses in the way in which user needs are being met; for example, by suppliers claiming to offer total solutions, but with weaknesses or absence of support or provision in some areas. Users and suppliers are also responding in the following ways:

- by forming joint ventures or consortia;

- by using systems integrators or managing agents to put a package together on behalf of a client

- by users reducing their needs to a level at which a single source can supply and guarantee a system – for example, one reflection of this is the market growth in smaller flexible manufacturing cells rather than large and complex systems

- by user firms carrying out configuration and project management in house on a 'do it yourself' basis

The supply side can be categorized into a number of different groups, as follows.

Computer suppliers

The main background of computer suppliers is of course in computers, but many have begun to diversify in a number of fields in recent years on the applications side. While computers have been used in the process industries for many years, their role in batch manufacturing has been primarily in data processing applications (such as payroll) and in production coordination: material requirements planning (MRP), computer-aided process planning (CAPP), and so on. Here new market possibilities are opening up in both new application areas (such as CAD, FMS, simulation and expert systems) and in networks and hierarchies of control which require central and distributed computer power. Some of these computer suppliers traditionally have strong links in industrial process computers or control instrumentation, which others have built up experience with microprocessor control systems which are much more widely used in manufacturing applications.

All of the major computer suppliers are offering CIM packages. These comprise computer hardware in a range of sizes and powers, and software – often of a modular nature-designed to integrate into a complete suite, with modules tailored for particular production management applications, networks, and so on. In the larger firms, such as IBM, diversification has gone further towards full factory automation supply capability, and recent acquisitions include CAD/CAM, robotics and factory automation networks.

As might be expected, many of these companies are themselves in the forefront of internal automation – often to an advanced level. Such investments have an important advantage in terms of learning by doing. Their experience can be used to improve the design of systems being sold and to sharpen the marketing approach taken because of greater understanding of the user's side of things. As a result they are becoming aware of the need to sell more than technology alone, if they are to provide customers with solutions to problems rather than just computer systems.

The main weaknesses of suppliers on the computer side are that they are often seen to be 'pushing boxes', without regard for the longer-term concerns of users, that they are trying to 'lock-in' customers to particular brands, that they are not experienced in all areas now related to CIM, and that they do not always offer the most appropriate solution (for example, offering high-cost and highly complex MRP2 systems when simple just-in-time approaches might be more suitable). There is also a high degree of cynicism amongst potential users of computer systems, who found many of the above problems in earlier generations of computers used in DP applications.

Software and systems houses

These are organizations primarily involved in developing software and designing systems, either on a freelance basis or as parts of larger firms. These organizations have the advantage that they tend to be independent of the hardware supplier (although some of their software is designed to run on particular machines) and they can configure semi-standard software to suit user needs. Several also offer new products – such as simulation packages – which are specific to the emerging needs of systems planners and developers.

The growth in integrated systems has been accompanied by expansion of this sector, often as the integration contractor in major CIM projects where software is seen as the key component. This has helped them to enter the market as systems integrators, and several are now offering a much broader range of services. Their main weakness is, however, that many software houses are still small operations and thus face problems in managing and implementing large-scale projects.

Equipment suppliers

These firms supply equipment, including machines, handling systems, robots and control systems, as components of CIM facilities. Increasingly, they have extended their range to include handling systems, such as pallet systems or simple robots, tool management systems, and so on. This pattern means that certain kinds of system can be offered on a turnkey package basis; for example, flexible manufacturing cells.

In other cases – such as on machinery for the food industry – the control systems may be much more specific. The trends are increasingly for such producers to offer more of a complete range of manufacturing systems and to negotiate licences or enter joint ventures in order to be able to do this.

The strength of this group of suppliers lies in their knowledge of their product and its traditional areas of application in batch manufacturing. Where user firms have traditionally bought single machine tools, they may well look for the same supplier to implement more complex manufacturing systems for them. Their weaknesses are their relative inexperience in selling whole systems, in project management, in organizational innovations (such as just-in-time) and in other aspects which a smaller firm might need, such as skills on the business side or long-term strategy. Many are learning by doing since their own product is made in small batches and to order; many also run user education and awareness raising programmes as part of their overall marketing approach.

Management consultants

A number of the larger management consultancies have begun to offer some form of automation and factory management consultancy as part of their overall portfolio. Many of these manufacturing-related operations have been around for some time and were originally concerned with projects such as work study, but they have now moved to offer expertise in AMT.

Their strength is in strategy and consulting. They are used to helping clients diagnose the need for and effect change in their organizations and to advising on strategic development. However, although these operations are growing quite large in some cases – reflecting both the expansion of the market and also the need for external advice – such consultancies are rarely able to provide the necessary resources themselves to design and build new facilities or to manage and commission large-scale projects. Their role is becoming more important as user firms look towards alternative and complementary innovations in the organizational sphere – such as just-in-time and total quality control –

in addition to major technological changes. Such innovation is well suited to consultancy since it requires skills in implementing organizational changes and is less resource intensive.

One other area of strength in the consultants' favour is their ability to arrange and manage some of the business aspects of AMT investment, such as arranging finance, taking advantage of government support schemes, and so on. This ability to provide a broad support service is of value to the smaller firm; although, as mentioned above, such breadth is not always matched by technological depth in terms of actual implementation.

Engineering consultants

These have much in common with management consultancies: the main difference is that their technical and production management expertise base is much broader because they have concentrated on engineering and factory projects as their main activity. Some engineering consultants have performed such roles as in-house groups in larger organizations. Others are organizations with considerable sector-specific knowledge which they are now trying to market, such as industry-specific research associations. Another important group involves those who originally entered the field selling microelectronics-related services and have now developed their skills and capabilities to address larger automation projects.

Systems integrators

This is a newly emerging group which aims at providing a measure of guarantee to potential users that the various pieces of technology and software chosen will actually fit and work together. The term is beginning to be used widely to cover many of the other categories mentioned above, but applies particularly to the kind of firm which can offer a strong track record of projects in which it has designed and built, managed and played other key roles. Systems integrators rely on a network of resources – from expertise providers, through computer hardware and software providers to equipment suppliers – on which a core team of highly skilled engineers can draw to configure solutions to meet particular clients' needs.

Early users

A final group are firms which entered the field early as users of integrated systems and are now selling on their experience. One reason for their doing this is that they were unable to find advice and support for the scale and type of automation project which they required, and this forced them into a 'do it yourself' solution. Their strength lies in the fact that they have succeeded, and can speak with first-hand experience, but a major weakness may be that they lack marketing experience in the field of automation products.

Skills and work organization

As we have seen in earlier chapters, the challenge posed by integration of technical elements is that similarly integrated skills are needed to support them. Increasingly, the trend is away from single skills tied to single functions, and toward multiskilling and flexible disposition of these skills. This in turn raises questions about patterns of work organization, and the inappropriateness of a system of organization which stresses division of labour and fragmentation of tasks, in a context where technology is essentially moving in the other direction, towards greater integration – a theme which will be picked up in detail in the following chapter.

Strategy

The notion of CIM as a strategic technology is not simply associated with its ability to provide a way of responding to the challenges of business in the environment of the 1990s. It also requires a long-term strategic plan to support its implementation. It cannot be bought off-the-shelf, as we have seen, but must be built up in a systematic fashion, gradually linking islands of automation together and building up a multi-level hierarchy. This requires reconsideration of the information needs and the control structures throughout the organization.

Developing such a strategic plan involves a high degree of top-level management commitment: CIM is not simply one more item of technology which can safely be left to the production engineers to specify and the accountants to justify. Firms which have succeeded in implementing integrated systems usually stress the considerable expense in time and resources which went into planning – and the converse is also true. One of the most common causes of failure is a lack of clear strategic objectives backed up by detailed planning.

Organizational structure and culture

The final barrier is probably the most difficult to overcome. Above all, CIM begins to challenge the accepted way in which organizations work on a day-to-day basis. As we have stressed throughout this book, it offers not just ways of improving what already is, but also opportunities for creating totally new ways of doing things. In systems theory it offers emergent properties, where the whole is greater than the sum of its parts. But these cannot be fully realized without significant organizational change – in structures, practices and, most importantly, in the underlying culture. It is these challenges which form the basis of the next chapter.

Notes

1   R. Boden and B. Dale, 'What is computer-integrated manufacturing?', *International Journal of Operations and Production Management*, 6 (3) (1986).
2   R. Kaplinsky, *Automation – the Technology and Society* (Longman, 1984).
3   For example, see J. Northcott, *Microelectronics in Industry* (Policy Studies Institute, London, 1987) or A. Majchrzak et al., 'Adoption and use of computerised manufacturing technology', in *Managing Technological Innovation* ed. D. Davis and associates, (Jossey-Bass, San Francisco, 1986).
4   Kaplinsky, *Automation*.
5   N. Garnett, 'Suspicious of a link-up', *Financial Times*, 4 January 1990.
6   H. Yamashina, K. Matsumoto and I. Inoue, 'Pre-requisites for implementing CIM – moving towards CIM in Japan', *Proceedings of Seminar on Computer-integrated Manufacturing*, Botevgrad, Bulgaria, 25–9 September, 1989, United Nations Economic Commission for Europe/IIASA, Geneva.
7   S. Parkinson and G. Avlonitis, *Proceedings of the 1st International Conference on Flexible Manufacturing Systems* (IFS Publications, Kempston, 1982).
8   R. Kaplan and T. Johnson, *Relevance Lost: the Rise and Fall of Management Accounting* (Harvard Business School Press, Boston, Massachusetts, 1987).
9   J. Sizer and G. Motteram, 'Costing of AMT at Rolls–Royce, Derby', in *Managing Advanced Manufacturing Technology*, ed. C. Voss (IFS Publications, Kempston, 1986).
10 P. Primrose, 'AMT investment and costing systems', *Management Accounting*, October 1988.
11 R Kaplan, Must CIM be justified by faith alone?, *Harvard Business Review*, March-April, 1986 p82-95.
12 Primrose ,'AMT investment and costing systems.'
13 'Looking for a payback', *Industrial Computing*, March 1988.
14 J. Wheeler, 'The evaluation of intangible benefits with the introduction of new technology', Centre for Business Research, Brighton Business School, Brighton Polytechnic, 1988 (mimeo).
15 Digital Equipment Corporation advertisement, 1985.

12 New Models for Old?

12.1 INTRODUCTION

In the preceding chapters we have been looking at the key technological elements in what many people believe is the next major wave of technological change to hit the world economy. Revolutionary shifts, such as occurred in earlier years with the emergence of steam power or electricity, also bring with them major challenges for social and managerial institutions. Radical technologies require nothing less than a new techno-economic paradigm.

As we have considered each technology it has become increasingly clear that several common themes are emerging. First, for all but the most simple substitution use of AMT, there is a need for significant adaptation, going beyond the normal organizational learning curve and finding new and radically extended answers to the new challenges that are being posed. Second, these changes are multidimensional: they do not just impact at one level or in one area but instead require the organization to change its entire system. This often involves a significant element of 'un-learning' – they require the organization to undergo a process of 'creative destruction'. Third – and most significantly – there is no natural or automatic change process associated with AMT. Simply buying the technology does not change the organization; that requires deliberate design and implementation effort. Failure to try to develop the two in parallel will lead to growing mismatch and the results will be counterproductive. This explains the inability of many users to get the best out of their investments, and also begins to suggest that there is a new form of 'best practice' blueprint emerging which involves radically different models for managing and organizing in manufacturing.

This chapter brings together these themes in an explicit form, focusing on the dimensions of change required to exploit AMT successfully, and the outlines of new emerging models for the factory of the future.

For convenience we will break this analysis into a number of organizational levels (although there are clearly many other ways of splitting this up). We will begin by examining the individual level and, in particular, the need for new skills and greater flexibility. Then we will look at group-level issues, largely within functional areas where the emphasis is on finding alternative patterns of work organization. Next we will turn to interfunctional and inter-group issues, and especially at the challenge of integration. At the level of the organization we will explore the question of control and the need for both greater integration and decentralization. The last level involves the question of inter-firm links and the extended model of the firm.

Central to all of this model shift is a new set of values and so we will conclude by reviewing some of the issues surrounding the complex concept of organizational culture.

12.2 ELEMENTS OF CHANGE

It is a feature of long waves that major technological changes bring into question existing arrangements for managing and organizing institutions – and as these become less appropriate, so a variety of alternative options are tried. While it is still unclear which of the many experiments now going on will lead to the new paradigm, there is at least sufficient evidence to point to some key underlying trends, which are summarized in table 12.1:

In each of these areas it is clear that there is no 'best' recipe for organizational design. Research has continually stressed the presence and importance of choice with regard to particular options. Although early researchers – for example, Joan Woodward in her pioneering work[1] – tried to find a direct, predictive relationship between technology and organizational structure, hoping that it would 'make it possible to plan organizational change simultaneous with technological change', this has proved to be impossible. The main problem is one of variety: different firms are confronted by different strategic contingencies, and choices (by managers and other key strategic actors) vary widely even in response to the same set of contingencies.

Instead, we are left with models which attempt to link choices and contingencies to provide at least an indication of broad trends. For example, Child[2] suggests the scheme shown in figure 12.1.

Using this kind of model we can suggest trends predicted by theory (such as the need for greater flexibility to cope with higher levels of environmental uncertainty) and to question the appropriateness of existing organization designs for newly emerging strategic or technological contingencies.

We can see that the (relatively) stable and predictable environment of the first half of this century has shifted to a highly uncertain and turbulent one. This requires a shift away from what Perrow[3] calls

Area	Past model	Emerging model
Skills	Single function	Multifunction
	Long skills life cycle	Short life cycle
	Skill life = employee working life	Skill life < employee working life
	Fixed relationship to tasks and equipment	Emphasis on flexibility and creative problem solving – need for understanding of whole system
	Stable skill/technology relationship	Variable skill/technology relationship
	One man, one skill demarcation	Cross-trading, multiskilling
	Training model of apprentice → craftsman	Continuing education
Work organization	High division of labour	Increasing integration of tasks
	One man, one job	Team working, role sharing
	Predetermined tasks and allocation	Semi-autonomous working groups
	Function/line-based	Cellular manufacturing
	Specialization and task fragmentation	Flexible teams, multiple roles and skilling
	Elimination of discretion	Increasing local discretion
	Payment by results	Alternative payment systems – for skills, for team output, for quality, and so on
	Supervisor as controller	Supervisor as resource
Functional integration	Specialization	Generalization/multi-functionality
	Differentiation	Integration
	Coordination through formal mechanisms, structures and procedure	Coordination through new roles, structures and informal mechanisms
	Demarcation and tight role specification	Loose role specification
	Boundary-limited	Boundary-crossing
Control	Control via formal rules and procedures	Control via informal mechanisms, self-organizing within broad framework

Table 12.1   Changing patterns of organization

Area	Past model	Emerging model
Control (cont)	Bureaucratic and mechanistic lines of control	Loose and flexible lines of control
	Pyramidal authority structure	Simultaneous centralization and decentralization (loose/tight)
	Vertical communication	Network communication
	Multi-level tall structures	Flat structures, few levels
	Emphasis on formal standards	Self-assessment
	Responsibility + power	Responsibility without power
Interorganizational relationships	Arms-length dealing	Close, cooperative links
	Short-term, transaction-based	Long-term, developmental
	Tight boundaries	Blurred boundaries, extensive crossing
	Tradition of confrontation with suppliers	Partnership in mutual development
	Lack of involvement of customer	Customer as an extension of factory
Culture	Mechanistic	Organic
	Rigid	Flexible
	Emphasis on local issues and goals	Emphasis on organization-wide issues and goals
	Non-participative	Participative
	Lack of ownership	Shared goals
	Single-loop learning	Double-loop – the 'learning organization'

Table 12.1   Changing patterns of organization (continued)

'routine' manufacturing towards 'non-routine' manufacturing, in which there is considerable variety and uncertainty in the tasks to be performed and an absence of previously prescribed and codified practice. In other words, we need a more flexible organization to deal with these contingencies. Organization theory gives us some guidance as to the kinds of design characteristics which are associated with the two types of manufacturing and, in the process, provides theoretical support for the kind of experimentation now taking place in a number of factories.

12.3 CHANGES IN SKILLS

In the area of skills, three broad changes can be identified. The first reflects a structural shift towards higher levels of skill required to support increasingly complex technologies. Data exists for a variety of industrial sectors and countries to confirm the general observation that there is a decline in the number of unskilled and semi-skilled workers in manufacturing, and an increase in the requirement for professional engineers, technicians and similarly qualified staff.

This is not only a function of the changing technological and employment pattern: it also reflects the changes in the educational expectations of individuals, who are now more conscious of the need for higher levels of qualification and who express a desire for greater personal development.

The second trend is towards greater convergence, with the shift from single skills towards multiple skills, required to support increasingly integrated and interdependent technologies. This process has been observed across a wide range of occupational groups and at all levels in the organization.

From a theoretical perspective we would expect this; for example, Thompson[4] suggests that more sophisticated arrangements involving 'sequential' or 'reciprocal' coupling (such as the inter-function linkages involved in integrated technology) will involve high levels of boundary crossing skills.

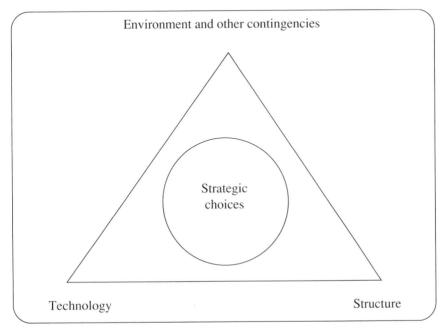

Figure 12.1   Child's model of organizational choice

The third trend is towards greater flexibility of deployment. Whereas traditional models involved clearly defined, predictable and predetermined tasks for which single skills could be developed, the emerging pattern is one which demands a more flexible response. The key skill here is the ability – or agility – to switch in different skills as and where relevant. For example, the task of maintenance is becoming more and more dependent on diagnostic and fault-finding skills, which require considerable flexibility and creativity, and correspondingly less so on actual repair skills.

An indicator of this is the emergence of new job titles and descriptions. For example, in the vocational education system in the Federal Republic of Germany a number of new '*Berufe*' (occupations) have been developed to reflect the skill needs in advanced manu-facturing.[5] Ettlie reports on similar patterns in the US.[6]

One consequence of this greater reliance on multiskilled and flexible support in AMT is the greater emphasis which it places, of necessity, on training and development. Whereas traditional models involved the acquisition of a single skill which would be applied throughout an employee's working life, the emerging pattern is one in which skills often have a short life cycle, and in which the skill in acquiring new skills and updating old ones becomes critical. Such 'learning to learn' depends principally on developing an understanding not only of 'know how' or ' know what to do' but also of 'know why' – a gradually deepening understanding of the broader context in which the worker is operating. This, in turn, moves the emphasis from task-related training to individual development, and from being a cost related to a particular piece of new technology towards being seen as an investment in organizational capability. Tacit knowledge and the ability to acquire and dispose skills in flexible fashion will become increasingly important factors in getting new technology to perform effectively, and many firms are now beginning to recognize the advantages offered by becoming a 'learning organization'.

The notion of strategic choice can clearly be seen in the case of skills. There is no longer any technological imperative which argues for a particular level or disposition of skill, but rather a range of choice open to managers. Whereas earlier systems emphasized deskilling and the removal of operator control and discretion, much of the current trend is in the reverse direction, in which human operators are seen as a key adjunct to successful implementation and flexible operation of advanced systems.[7]

12.4 FLEXIBLE WORKING PRACTICES

A key feature, highlighted in chapter 5, has been the shift towards more flexibility in the patterns of work and of skill distribution. Atkinson identifies three major elements in such a model of 'the flexible firm':[8]

1 Functional flexibility: the ease with which tasks performed by workers can be adjusted to meet changes in technology, markets or other contingencies. This relates strongly to the preceding discussion on multiple skills which provide the opportunity for shifting workers around in response to these uncertainties. Low levels of skill or lack of multivalency in skill can inhibit flexibility.

2 Numerical flexibility: the ease with which the number of workers can be adjusted to meet fluctuations in demand or other contingencies. Here the main strategy used is to operate with a core plus a peripheral workforce, the latter component of which is retained on short-term, part-time contract and can be expanded or contracted rapidly. Atkinson estimates that between 5 and 6 per cent of the UK workforce are engaged in temporary work of this kind. Alternatives to this core–periphery model for numerical flexibility exist for dealing with fluctuations of demand – (such as multifunctional workers who can be retained within the core but shifted around the factory to where the work is).

3 Financial flexibility: the extent to which the structure of pay encourages and supports numerical and functional flexibility. Here the challenge is to find alternative and less rigid payment systems which provide encouragement to operate in more flexible modes; for example, many firms are experimenting with schemes which reward the acquisition of additional skills.

The emerging pattern appears to be that capital-intensive firms place greater emphasis on securing functional flexibility (in order to keep utilization levels high), whereas labour-intensive operations (such as assembly work) are characterized by emphasis on financial and numerical flexibility.[9]

Functional flexibility is achieved through various kinds of broadening mechanism – cross-trading, multiskilling, second or dual skilling, and so on. There is growing evidence of this pattern, especially at the technician and craft levels; for example, Fleck reports on the emergence of a new job category concerned with robot maintenance, which combines skills in electrics, electronics, mechanical engineering, hydraulics and pneumatics.[10] Similar findings are reported by Cross and Mitchell[11] for the process industries: they suggest that existing skills should be regrouped into five categories – craft technician, system specialist craftsmen, machine specialist craftsmen, dual-trade craftsmen and cross-traded craftsmen – and two new categories, of 'mechanicians' (with skills in diagnosis and fault-finding), and 'user-maintainers '(which brings maintenance skills back into the hands of the operators). Senker[12] reports the same kind of pattern for a range of occupations within the engineering sector. Such flexibility is not confined to the shop floor; amongst graduate engineers there has been a shift towards more broadly based training and qualification, such as the manufacturing systems engineer concept advocated by Parnaby.[13]

Moves of this kind may appear to be technologically necessary, but they often challenge the basic structure of organization within old-established plants. For example, in the Shell Chemicals plant at Carrington, Manchester a new agreement has abolished the original structure consisting of 14 craft trade groups (such as laggers, welders, plumbers, and so on) and replaced them with a single grade. All non-management personnel, including clerical and catering staff, are now classified as 'technicians' and the pay system has been massively simplified. In order to provide the necessary skill mix extensive redesign of jobs took place, and a heavy investment has been made in training so that every plant technician will be sufficiently skilled to handle about 80 per cent of traditional craftsmen's work across all traditional trades.

This change towards more flexibility requires a major strategic commitment to change on the part of the organization. As the personnel manager explained:

> . . . we'll manage flexibility in future . . . we are determined not to slip back into allowing people simply to do what they feel comfortable doing. It's one bargain, one code of working and everybody does everything which their skills and knowledge allow them to do safely. Cultural and attitude changes are required.[14]

A similar approach has been taken at the Darlington, UK plant of Cummins Engines, where an extensive flexible manufacturing facility has been installed. This has been accompanied by a reduction in the number of job grades, a change in the payment system to reflect reward for skill acquisition, and the development of a flexible working culture.[15]

Numerical flexibility is increasingly concerned with alternative working hours and the use of part-time peripheral workers supporting a full-time core. Here the part-time element may or may not be employees of the firm: they may also be self-employed or subcontractors. A number of plants have begun to adopt alternative working hours, with some notable experiments that break with tradition. For example, in the new BMW plant in Regensburg, West Germany, the heavy investment (£500 million) in advanced flexible manufacturing technology requires a high level of utilization. Accordingly, the firm has moved to a four-day, nine-hour shift, 35-hour working week, in which each working place is effectively 'manned' by 1.5 people. This means that the factory itself runs for six days a week, giving 54 hours of utilization (as opposed to the usual 40 hours) with an 'up-time' of around 95 per cent at present. Workers work four days and then have three days off; in addition, their attendance is planned so that every two weeks they have a five-day weekend.[16]

It seems as though similar patterns will become increasingly common as firms seek to maximize the utilization of expensive capital plant. Ford of Europe recently announced its intention to move to a 24-hour, six- or seven-day production system, following General Motors' introduction of a three-shift system in Antwerp. Toyota, which is investing in a new plant in Derby, UK, has also indicated its plans to adopt 24-hour, six days a week working. These changing shift patterns will be accompanied, in the Ford case, by the use of semi-autonomous working groups to achieve higher levels of functional flexibility.[17]

Financial flexibility is seen by many firms as essential to ensure the changeover to more flexible practices. As in the examples of Shell and Cummins above, it is often associated with restructuring, simplifying pay scales and offering new forms of 'payment for flexibility'. A similar point is made in a recent study of AMT introduction in the US: '. . . paying employees according to their mastery of a progressively broader range of tasks is often adopted to encourage both the learning and flexibility in work assignments required in the operation of AMT'.[18]

12.5 WORK ORGANIZATION

In the context of new technology, patterns of work organization can be related to three key contingencies, according to Cummings and Blumberg[19]. These are:

- technological interdependence – the extent to which the technology requires co-operation among employees to produce a product or service

- technological uncertainty – the amount of information processing and decision-making which employees must carry out

- environmental stability

Put simply, as systems become more complex and interdependent (as in modern manufacturing technology) and as environments become more turbulent and less predictable, so there will be an increasing need for flexibility. The requirement for rapid response focuses attention on models of work organization which permit a degree of local autonomy and self-organization, a high level of functional flexibility and group or team rather than individual working. These represent a shift away from more conventional approaches which emphasize functional or line arrangements and procedural control.

The developments which we have seen towards CIM provide strong support for this trend. There is growing interest and experiment in areas such as group technology and cellular manufacturing, and a resurgence of interest in the principles of socio-technical systems design. Typically, in such models workers are given a broad task specification and within those loose constraints are allowed to organize, to allocate roles, to schedule tasks, and so on. Functional flexibility is being sought (especially in capital-intensive industries) and this is allied to greater functional integration, bringing many specialist functional roles within the manufacturing cell – such as maintenance or quality control. The emergent picture is of the ' factory within a factory'.

The role of supervision is also changing as a result of devolution of some power and control to these units. Strategies based on what Friedman calls 'responsible autonomy' are increasingly being applied, and these shift the emphasis of the supervisor's role from that of police-man to that of facilitator.[20]

Once again, the issue of strategic choice is central to this discussion. There have been a wide range of studies of new manufacturing technologies, both in substitution and more integrated form. They all support the view that there is no connection between the technology and the form of work organization adopted, at least not in the sense of a technologically determined 'best' arrangement. Rather, it is a case of achieving the most appropriate fit between technology and organization. A variety of options exist: for example, in their studies of CNC machine tools in the Federal Republic of Germany, Rempp found no less than five different and economically viable patterns of work organization. This point is reinforced for CAD, for FMS and for a variety of other applications.

This focuses attention on the question of how the strategic choices are made, and the evidence suggests that this is very much a matter for negotiation between the relevant actors in the process. Although managers represent key actors the process is , as Wilkinson points out, a political one in which other interest groups are involved. It is useful to consider it taking place within a 'design space', an area which defines the room to manouevre and which is shaped by a variety of factors.[21]

In the context of advanced and integrated systems it is clear that the same picture of choice exists – and this is to be expected at the stage of experimentation in the emergence of a new paradigm. Simultaneously, there are both attempts to find new arrangements and to apply old and less appropriate ones – a point brought out by Scott and Jones in their studies of FMS in the UK and US. Here they found not only examples of 'post-Fordist' approaches , in which completely new arrangements were being sought, but also what they term ' neo-Fordist' and 'neo-Taylorist' arrangements, which carry with them some of the elements of the old paradigm.[22]

It is important to note that the options available in work organization design are necessarily dependent on the available skills base in the organization. Where levels of skill are high, or where there has been training to support multifunctionality, then different options are open. This may explain why experiments in team working and in the use of multiskilled workers are more prevalent in Scandinavia and in Japan than in the UK or the US, where the relative skill levels are much lower and are distributed more narrowly across the workforce.[23] As Campbell and Warner comment, quoting Peter Senker, '. . . years of neglect by management have resulted in skills shortages which in turn place constraints on the range of strategic options open to management regarding new technologies and products.'[24]

Our own research on flexible manufacturing systems in Sweden and the UK provides one example of this: the move towards flexible man–machine cells is much more developed in Sweden than in the UK where many firms seek to use technology to replace or substitute for skilled labour.[25]

One feature of such moves towards alternative work organization designs is the shift in payment and incentive systems. Whereas in the Fordist model the theme was payment by results and focused strongly on the individual, the new pattern requires more emphasis to be placed on factors other than output (for example, payment for skills acquisition, for quality, for group productivity, and so on). In order to facilitate flexibility, many of the complex systems of demarcation and pay differentials have to be removed and career progressions need to be developed that are more in line with those of staff. Systems are needed which aim to build some sense of 'ownership' of the problems and of the organization: these range from share ownership schemes through a host of other approaches to employee involvement.

12.6   FUNCTIONAL INTEGRATION

At this inter-group level the dominant issue is the extent to which increasing technological integration forces functional groups to work more closely together. For example, CAD/CAM systems require a different and much closer relationship between design and manufacturing. Computer-aided production management systems redefine the information-sharing relationships between purchasing, marketing, production control and other groups. Flexible manufacturing systems require closer links with quality and maintenance.

Environmental uncertainty is also a strong force in this trend towards greater functional integration. As Lawrence and Lorsch noted, back in the 1960s, the greater the environmental uncertainty, the more the need for closer coupling between functions to reduce the timespan of feedback to that environment. They also made the important point that different modes of integration were more or less effective under different circum-stances. For the kinds of flexibility and rapid response demanded by today's manufacturing environment it seems likely that coordination using formal procedures may be less effective than less formal loose arrangements.

Flexible forms of coordination of this kind approximate to what Mintzberg terms 'ad hocracy', and they can be achieved in a variety of ways. Much stress has been placed on the idea of matrix management, job rotation and project team structures and – as Ettlie and others have pointed out – there are in practice a wide range of experiments taking place, all aimed at developing closer linkages between functions. The process of experimentation once again suggests the search for alternatives for the new paradigm.

Child[26] suggests that organization design offers two basic approaches to dealing with this question:

- integrate through convergence within the same role tasks which were previously performed within separate roles

- integrate through closer coupling or overlapping of different roles

The former approach is associated with things such as the blurring of roles within teams or the creation of new, hybrid roles. This links back to the earlier discussion of multiskilling or multivalent working, and also raises the question of the need for both specialists and generalists. Examples of the second type of approach include various structural mechanisms such as matrix and project team approaches.[27]

12.7 VERTICAL INTEGRATION AND CONTROL

In this area we would expect that increasing environmental uncertainty would require more decentralized decision-making. Control shifts away from a procedural and planned basis – doing things by the book accord-ing to predetermined rules – and becomes instead a much more locally determined activity, operating with autonomy within broadly agreed and predefined limits. This in turn changes the traditional process of monitoring and control, from direct supervision to models based on local 'responsible autonomy'.

In Thompson's terms, highly interdependent systems designed to deal with high levels of uncertainty require decentralization of control down to the point of action – the 'sharp end'.[28] By contrast, 'sequential' activities (in which workflow follows particular stages serially, in a planned and predictable fashion) and 'pooled' activities (in which there is relative independence of operating units), are both more amenable to centralized planning and control.

Once again the question of choice emerges here. On the one hand there are many who see new technological systems reinforcing the traditional paradigm, increasing centralization and concentration of control. On the other are a variety of experiments which seek to decentralize and to evolve alternative patterns of communication and control which support greater local autonomy. As Campbell and Warner point out, there is a growing paradox in that the technology accentuates the range of choice: '. . . with each successive development, the greater the potential for organisational decentralisation, the greater the potential for organisational recentralisation'.[29]

The nature of computer-integrated manufacturing illustrates this, with its emphasis on centralizing information in a common database on which all functions can draw. This inevitably offers opportunities for increasing contration of power and control – a point made by commentators, such as Perrow[30] Child suggests three ways in which AMT can extend the possibilities for greater control:

● AMT systems offer faster, more comprehensive information availability about the state of operations throughout the business

● decision support and sophisticated analytical tools eliminate dependence on evaluative judgements and reports from supervisors and middle managers – the number of control layers can thus be reduced

● the trend to integration includes an integration of the control system so that a total organizational picture is available – the factory behaves like one complex machine rather than as a collection of several units

The counter-argument – for greater decentralization – rests on the need to bring decision-making autonomy closer to the operating units in order to deal with increasing environmental uncertainty. Flexible reponse can be offered through a combination of flexible technology and organization. One indicator of this approach is the growth of concepts such as 'factories within factories'. Morgan[31] discusses the idea of 'holographic organization design' in which all the elements of the whole organization are reproduced in its parts, much as in the technology of image capture via holography.

It is possible to see a resolution of these apparently contradictory trends within the technology itself. Early generations of computer control were highly centralized, depending as they did on mainframe computers. Communication and control followed the organizational structure of hierarchical authority. Later generations made available greater local power – through the use of distributed systems – but the emphasis was still on hierarchical control, with those distributed terminals and smaller computers connected directly to the central mainframe. However, current technology is predominantly about *networks*, the essential property of which is that they can be both highly centralized and completely decentralized at the same time.

Thus it is possible in a CIM facility for the top level of management to have a central overview of what is taking place throughout the business, while at the same time local units can operate independently or in collaboration with each other with only minimal reference to the centre. Such models again correspond to Mintzberg's 'ad hocracy' and to what Peters and Waterman term 'simultaneous loose/tight properties'.

12.8   INTER-ORGANIZATIONAL INTEGRATION

As we have already noted, the trend to integration in technology does not stop at the boundary of the firm, but extends into the supply and distribution chains. AMT facilitates a variety of inter-organizational linkages, such as shared CAD and electronic data interchange throughout the distribution chain. The emergence of such networks is characteristic of the newly emerging paradigm, and there is growing evidence of its viability as an alternative model of economic organization. Indeed, much of the success of the just-in-time system is predicated on alternative supplier relationships, while many commentators also draw attention to the more tightly networked arrangements in Japan involving manufacturing and finance.

In theoretical terms, uncertain and turbulent environmental conditions are forcing firms into relationships in which coupling shifts from stable 'pooled' arrangements, in which each unit is largely independent of others, towards 'reciprocal' modes which require much higher levels of interdependence. The consequence is that inter-organizational coordination arrangements shift from traditional links governed by procedure and rules to more fluid and co-operative arrangements which stress longer-term partnerships and mutual development. This in turn represents a major challenge, as Lamming points out, to the basic type of adversarial relationship which has existed in many industries.[32]

Links between firms take many forms, but the implication of the new model is essentially one of gradually growing together towards what some commentators call 'co-makership'.

But the challenge is, once again, to the relationships between firms; for as long as the dominant culture is adversarial, and based on power rather than co-operation and mutual development, then the necessary openness on information and the long-term shared strategic perspective will be lacking, and will inhibit the effectiveness of such networks.[33]

12.9   CULTURE

Central to the idea of a 'paradigm shift' is the emergence of a new set of values and norms. We can see this pattern clearly if we examine recent evidence on the implementation of major technological change, where the necessary organizational changes have been accompanied by radical shifts in organizational culture. Conversely, many of the failures can be related to trying to locate new ideas (technological or organizational) within an old-established and unchallenged culture.

It is difficult to get to grips with the concept of culture, but its pervasiveness makes it the most important area which has to change if technological innovation is to succeed. A working definition would be the set of norms, attitudes, values and beliefs which determine ' the way things are done around here', the way in which the organization views its world and reacts to challenges within it. It incorporates the common 'taken for granted assumptions' which inform everyday life in the factory.[34]

Discussion of the relationship between culture and operating environment can be traced back at least to Burns and Stalker,[35] who pointed out the two different types associated with stable and changing activities. 'Mechanistic cultures' were typical of stable environments such as production, where activity is predictable and it is possible to do things by the book. Amongst the key characteristics of such organizations were:

- specialized differentiation of functional tasks, in which problems facing the firm are broken down into small elements
- the abstract nature of each individual task, which is carried out with techniques and purposes more or less distinct from those of the organization as a whole
- reconciliation of the performance of these different tasks into those of the organization through a hierarchy of superiors
- tightly and precisely defined functional roles, in terms of rights, obligations, technical methods, and so on
- control via the hierarchy
- a hierarchical structure reinforced by the exclusive restriction of knowledge about what is going on to the top of the pyramid
- interaction tending to be between superior and subordinate
- operations and working behaviour governed by superiors
- loyalty and obedience valued
- greater importance and prestige attached to internal (local) knowledge rather than general (cosmopolitan) knowledge, experience and skill – an inward-looking orientation.

By contrast, innovative activity (such as in R & D) was characterized by high levels of uncertainty and thus required a different culture, which they termed 'organismic'. Its characteristics include:

- the contributive nature of special knowledge and experience to the common task of the concern

- the realistic nature of the individual task, which is seen as set by the total situation of the concern

- the adjustment and continual redefinition of individual tasks through interaction with others

- the shedding of responsibility as a limited field of rights, obligations and methods – problems may not be posted upwards, downwards or sideways

- the spread of commitment to the concern beyond purely technical definition

- a network structure of control authority and communication

- omniscience not imputed to the head of the firm – knowledge may be located anywhere in the network and this location becomes the centre of authority for that aspect of the concern's activity

- a lateral rather than vertical pattern of communication

- a content of communication which consists of information and advice, rather than instructions and decisions

- commitment to the concern's tasks and to its progress, and the respect for this rather than for blind obedience or loyalty

- importance and prestige attached to affiliations and expertise which has validity outside the narrow environment of the firm – an outward orientation

Clearly, the pattern of environmental change which we have been discussing is forcing many firms to move from stable mechanistic approaches to more fluid and flexible organic ones. Handy uses the terms 'role culture' and 'task culture' to describe mechanistic and organic organizations and suggests that 'change in the environment requires a culture that is sensitive, adatable and quick to respond. A task culture is most suited to coping with changes in the market or the product . . . diversity in the environment requires a diversified structure. Diversity inclines towards a task culture.' He also introduces another model, the 'power culture', which is primarily a model associated with small entrepreneurial organizations and characterized by networks with a key power source at the centre. Such cultures have many of the adaptability advantages of task cultures but depend on the decisions or whims of a core group or a charismatic individual at the centre.

Perrow focused particularly on the link between culture and technology in organizations, and developed the ideas of Woodward and others who were trying to find a relationship between technology and structure. He argued that there was a continuum in technology running from routine to non-routine operations and explored the impact of technology on a wide range of elements, including the systems of coordination, control and interfunctional operation, as well as on more individual-centred factors such as motivation. Drawing on work by Child[36] and Handy[37] it is possible to relate some of the design and cultural features which we might expect from these different poles on Perrow's continuum, shown in table 12.2.

Routine manufacturing	Non-routine manufacturing
Highly departmentalized and functional structure	Flexible, geared to rapid response Functional integration, blurring of boundaries
High percentage of managerial, administrative and staff employees	Skill-based local autonomy – high levels of discretion in supervisors and team
Heavy reliance on formal procedure for planning and operational control	Control through responsible autonomy and group planning within broad limits
Integration through impersonal modes – for example, setting work programmes	Integration through feedback (mutual adjustment) and informal mechanisms
Emphasis on information and control systems reporting variance from pre-determined standards	Self-regulating, responsibility for standard-setting and maintenance
Centralized policy initiatives with little delegation	Local and devolved autonomy – few hierarchical levels
Suited to predictable, stable environments	Suited to uncertain environments
Suited to high-volume, low-variety production	Suited to low-volume, high-variety production
Role culture is appropriate	Task or power culture is appropriate

Source: derived from Child (1987) and from Handy

Table 12.2   Routine and non-routine manufacturing

A central feature of current discussion of changes in organizational culture is the need for shared values and goals to which all organization members are committed. Such an 'everybody on board' approach, characterized by a sense of ownerhip and participation is, again, at odds with the traditional pattern which emphasised the marginalization of workers and instrumentality as motivation. Implicit in such models is also a challenge to the traditional allegiance to the goals and values of sub-units and functional groups. The move to integrated systems requires a more integrated approach on the part of the organization, and this cannot be achieved simply through structural changes. It requires shared values and beliefs and a sense of common purpose.

As we noted in chapter 6, this also poses a challenge for organizational learning. Organizations need to move beyond their traditional models of learning, which are designed to improve the way things are done without challenging the underlying norms or values. Such single-loop approaches are appropriate in stable conditions, but where uncertainty is high and where the demand is for greater participation, the need is for approaches which encourage challenge and continuous improvement.[38] Argyris and Schon term this 'double-loop learning' and argue that 'this occurs when an error is detected and corrected in ways that involve the modification of an organisation's underlying norms, policies and objectives'.

Considerable support can be found for the need to change approaches to learning. Crozier, for example, in his critique of bureaucracy, points out that a bureaucratic organization is one which 'cannot correct its behaviour by learning from its errors'.[39] Mintzberg suggests three models of strategy, moving from seeing it as an essentially mechanistic process to one which implies a much more organic approach: strategy as rational plan, strategy as vision and strategy as fluidity and learning.[40] Much discussion is now focussed on the concept of 'the learning organization',[41] and in a recent report the OECD placed particular emphasis on the need for such a shift.[42]

There is also growing emphasis on organizational learning as a source of competitive advantage at the level of the firm. For example, Ettlie quotes Buss and Guiles discussing the GM Saturn project, in which there has been considerable learning both from problems in the high technology area and from experiences elsewhere, particularly the NUMMI joint venture with Toyota: 'Saturn planners now are emphasizing the more efficient organisation of human workers, learning from the success of that approach at GM's California venture with Toyota Corporation'.[43]

12.10   EXPERIMENTS TOWARDS A NEW MODEL

Although we can, as in the above discussion, outline some of the dimensions of organizational change , there is no blueprint available for the newly emerging paradigm. Instead we have a pattern in which the traditional Fordist approach is seen to be less appropriate and where a variety of experiments are taking place to try and find alternatives, some of which may be viable in the long term. We will look briefly at some of these.

For example, flexibility is being sought through multiple full-time job assignments which Ettlie describes as 'like job-sharing with yourself'. It is used as an alternative to job rotation between functions or departments, especially in support of 'simultaneous engineering' projects. In this model an engineer might work on job A from eight till lunch and on job B from lunch till the end of the shift, perhaps in a different plant and reporting to a different manager. Simmonds[44] comments that this pattern has long been present in the marine industry, where engine rooms in ships have run on the basis of multiskilled, multifunctional engineers who work in teams and rotate tasks as a means of acquiring new skills and personal development.

At the Exxon ethylene plant in Mossmoran, Fife, a similar scheme has been introduced whereby technicians work 12-hour shifts for 16 weeks and then switch to day shifts for eight weeks, carrying out maintenance work. This scheme was introduced to try to break down barriers between craft and process areas, and has had the additional benefit of improving maintenance and reducing downtime.[45]

A variety of moves have been made towards multiskilling, with particular emphasis on operators carrying out quality and maintenance work. JIT and TQC programmes inevitably move firms along this path. For example, Coca-Cola–Schweppes have introduced a highly automated production line at their Wakefield plant in the UK. This £60 million facility is designed to operate at line speeds of around 2000 bottles per minute, 25 per cent faster than competitors, and to run for 24 hours per day and seven days per week. In order to support this they have introduced team working, with one team responsible for two lines, and within the teams have moved to a multiskill pattern which ensures much greater flexibility.

Similar concern for keeping expensive plant operating through high worker flexibility can be seen in the car industry; for example, Volkswagen have been using a grade of operator called an '*Anlagenführer*' (essentially a monitor of complex equipment) since the 1960s, and this role has been extensively deployed in recent plants. This pattern is, of course, more familiar in the process industries where refineries and other installations have used this model extensively.

To support these moves firms are trying experimental payment systems – payment for skills, group work and multiple job assignments. Schemes of this kind do not always work: Ettlie reports that a mid-Western plant using FMS planned to reward operators on the basis of system uptime (the time available for production). The programme failed because the remainder of the factory wanted similar arrangements – emphasizing the need to make *systemic* rather than local changes. Another payment-related feature which is emerging is employee share ownership schemes (ESOPS), extending the alternative payment scheme approach to engender a sense of 'ownership' amongst employees.

Firms are also beginning to use performance appraisal as an alternative to traditional hourly-paid performance measurement and reward system design. Ettlie observes that these often lead to groups of workers who traditionally ignore organizational change programmes becoming involved and motivated, with consequently beneficial results, such as suggesting technology improvements.[46]

Alongside changes in payment systems are moves towards establishing single-status employees. Here the intention is to eliminate much of the demarcation problem and to simplify pay scales and increase flexibility. Examples include the Cummins plant discussed above, and the Westinghouse Nuclear Fuel plant in Columbia, South Carolina (Graulty et al., cited in Ettlie). Here the traditional grades have been replaced with a single 'team member' grade. Part of this programme has been the agreement of a shared statement regarding the goals of the project which stresses maximum productivity and employee participation. The system which was developed aimed to maximize flexibility, promote skills acquisition and promote co-operation and team work. In Ford UK attempts have been made to reduce the number of job classifications and since 1985 these have been reduced from 550 to 52, with a comparable increase in the range of tasks carried out by operators.[47] In the UK there are now only two distinct groups (electrical and mechanical) while in Germany there is only one grade, the *'Industriemechaniker'*, who is essentially a 'jack of all trades' and capable of a wide range of tasks in support of AMT.[48]

The changing pattern is typified in the experience and plans of ICI, the giant UK chemicals manufacturer. They employ around 29 000 manual workers on 60 sites and have recognized the need for radical changes including multiskilling and more flexible work patterns. In proposing a major and continuous long-term change programme, the Personnel Director, Derek Holbrook, comments:[49]

> . . . multi-skilling, more flexible working practices, teamwork are all too definite and limited . . . what we are talking about is the legacy of the past which surrounds what has traditionally been called manual work . . . we do not want a few adjustments to demarcation lines or a specific series of tightly negotiated changes which may form tomorrow's restrictions. We want a significant shift in the whole legacy of the traditional shopfloor, from the way people work, to their attitude to the company and relations with managers.

Interfunctional co-operation can also be seen in the increasing use of teams to design, develop and implement new products across an organization, carrying all relevant personnel throughout the process rather than on a sequential handover basis between functional representatives. Work organization changes emphasise a high degree of restructuring around teams – essentially building upon the established principles of socio-technical systems design. Wall and Clegg[50] report on several programmes which involve autonomous working groups with positive benefits for system performance; measured, for example, as up-time of systems. The involvement of the workforce in incremental development becomes critical to successful plant operation; amongst examples of this kind they cite the case of Flygt, the Swedish pump manufacturer, where the work groups within flexible assembly cells played an important role in developing the plant. This experience was also common in our Swedish FMS study where considerable devolution of responsibility led to significant improvements in organizational flexibility.

When the Japanese tyre maker Sumitomo took over the ailing Dunlop concern in the UK it introduced a number of changes, including group working. SP Tyres, as the company is now called, reduced the number of hierarchical levels in the plant and created small cells with direct responsibility for all aspects of production. The results have been impressive: output has risen by 50 per cent since 1984 despite a 30 per cent reduction in the workforce, quality has improved by 50 per cent and market share has been increased, turning a loss of £20 million to a 1988 profit of £3.7 million. Although some £10 million per year has been invested in the facility, little of the improvement is attributable to the sweeping introduction of new technology but rather to these changed practices.[51]

Another emerging feature is emphasis on team and organizational development. Here the intention is to improve the effective working of teams by developing skills (such as open-ended problem-solving) and also to resolve conflicts between team members. It aims to maximize flexibility and capability within groups, by building on effective group processes and developing individual and group interpersonal skills. Henderson and Horsley[52] report on an application in the ICI Agricultural Division, in which the problems and delays associated with commissioning major methanol plants were dramatically improved through the adoption of such an approach.

In another case, the UK sweet maker, Trebor, has been applying organizational development principles in its plants for many years. Its workforce is now structured into loose teams and emphasizes transferable skills. In one plant, for example, there are 13 work groups, responsible for everything from production to catering, engineering and employee welfare. Each group contains between 5 and 12 people who organise themselves in terms of job rotation, timing, quality, and so on, and who are also responsible for non-routine activities including recruitment and individual training and development.[53]

With the trend to manufacturing cells, opportunities are opened up for self-organizing groups which can allocate roles, schedule activities, and so on. This provides for higher levels of individual motivation, improves intra-group communication and in many cases ensures higher productivity and quality. Such arrangements are often noted in just-in-time programmes (for example, Medina and Cartaya report on a Venezuelan shoe factory where such a shift had marked benefits for the firm[54]), although their genesis can be traced back to the early days of socio-technical systems design in the coal and textile industries (Miller and Rice[55]) and to work in plants such as the Volvo Kalmar facility in Sweden. Lessons learned in the Scandinavian experience have been applied to an ambitious new programme involving some 40 major manufacturers, the results of which are typified in the new Uddevalla plant in Sweden for the Volvo company.[56] In this the lessons of Kalmar are taken further in an attempt to blend the best in 'high technology in engineering' with ' true friendship and co-operation'.[57] The teams (in a plant in which, incidentally, 40 per cent of the workforce is female) decide how many cars will be assembled each day and final assembly is done in place with a team of five or six people. Teams rotate from pre-assembly to final assembly. Although it is still early in its life, the absenteeism level (a major problem in Swedish industry) for the Uddevalla plant is only 8 per cent, compared with levels of 17–25 per cent at the larger Gothenburg plants. Saab is experimenting with a similar facility in its Malmo plant, and other experiments are taking place elsewhere in Scandinavia.

Another example can be seen in the NUMMI joint venture between Toyota and General Motors in the US.[58] The New United Motors Manufacturing Company opened in late 1984, using a plant originally closed by GM in 1982 in Fremont, California. Toyota invested $150 million and GM $20 million plus the plant. The facility employs around 2500 workers on two shifts per day production and output runs at about 950 cars per day. GM are responsible for marketing and Toyota for production management and sourcing.

Although no new investments have been made in AMT the improvements in performance of the plant over its predecessor are dramatic. Where absenteeism was as high as 20 per cent it is now less than 3 per cent, and productivity is high (at around 22 man–hours per car) in comparison with other US car plants. The main changes have been in organization, where extensive use is made of multiskilled teams, reduced job categories (three instead of the original 200) and reduced layers in the hierarchy (down from nine to four levels of management).

These new approaches are not always successful. Ettlie reports on the case of the GM Coldwater facilty in the US, where such an approach involved the UAW and company working together to make dramatic changes in work content through job rotation as they introduced robotics. Unfortunately, senior management resistance to this new 'jointness' was a major problem which persisted and may even have contributed to the eventual closure of the plant.

In the Scandinavian case, despite extensive attempts to improve the quality of working life, there is still a major problem with absenteeism and with a reluctance on the part of young people to work in industry.[59] For example, Swedish absenteeism per worker is running at 29 days per worker per annum, in comparison with 11 in the UK and 18 in West Germany (in 1989). Rates in some companies are excessive; for example, Volvo reports 14 per cent, Asea–Brown Boveri 18 per cent, Electrolux 15 per cent and the Trelleborg Group 21 per cent. Faced with this challenge, the one clear message for manufacturers is that even more effort needs to be placed in ensuring the right kind of working environment, one which is challenging and offers responsibility and opportunities for personal development. It is this which underpins the recent moves towards more radical factory structures such as those described above.

Emphasis is also being placed upon devolution into minifactories or manufacturing cells or, at a higher level, into 'strategic business units'. Such a focused approach, concentrating production of one family in a minifactory, can be seen in the new Volvo Uddevalla plant, for example, where small groups of 80 workers run a series of parallel miniplants. At the same time there are moves towards a reduction in levels in the hierarchy and with it a changing role for many managers and supervisors. For example, Ford of Europe are now moving towards a model which has group leaders responsible for teams of 8–12 staff plus 'area foremen' who are responsible for areas of the plant, often covering 25–30 staff. In Volvo's Uddevalla plant, all planning and control is devolved down to groups of ten, with no direct supervision.

12.11 SUMMARY

From the foregoing we can begin to see what a 'new paradigm' organization might look like. The outline is still hazy and its boundaries are changing, but the vision is becoming clearer as more firms experiment, and learn from both successes and failures. It is interesting to note how – just as in earlier paradigms – there was a clear geographical focus to the emergence of new models: the experience of Japan seems to be closest to the newly emerging pattern. Most of the lessons we have learned in recent years about Japanese management and organization underline principles at the core of the new paradigm – continuous improvement, employee involvement, networking, functional integration, and so on. But it is also important to reflect on the fact that earlier paradigms may have begun life in a particular geographical area but eventually diffused on a worldwide basis. There is no Japanese patent on the new paradigm: it is something which can be – and is being – adopted by firms in any country, and adapted and developed to advantage.

The key requirement for entering the new paradigm is the willingness to challenge and change the ways in which things have traditionally been done and to question the underlying values. Such a move towards becoming a learning organization is not easily accomplished, and the next chapter explores briefly some of the challenges involved in the implementation of change on this scale.

Notes

1 J. Woodward, *Industry and Organisation* (Oxford University Press, 1970).
2 J. Child, *Organisations; a Guide to Problems and Practice* (Harper and Row, London, 1977).
3 C. Perrow, *Organisational Analysis: a Sociological View* (Tavistock, London, 1970).
4 J. Thompson, *Organisations in Action* (McGraw-Hill, New York, 1967).
5 D. Buschhaus, *Vocational Qualifications and Flexible Production as Demonstrated in the Industrial Metalworking and Electrical Engineering Occupations* (Federal Institute for Vocational Education, Berlin, 1985).
6 J. Ettlie, *Taking Charge of Manufacturing* (Jossey-Bass, San Francisco, 1988).
7 M. Warner, W. Wobbe and P. Brödner, (eds), *New Technology and Manufacturing Management* (John Wiley, Chichester, 1990).
8 J. Atkinson, *The Flexible Firm* (Institute of Manpower Studies, Sussex University, 1984).
9 J. Tidd, *Flexible Manufacturing and International Competitiveness* (Frances Pinter, London, 1990), and S. Rosenthal and M. Graham 'Flexible manufacturing systems requires flexible people,' *Human Systems Management* 6 (1986), pp.211–22 provide reviews of this experience in the UK and US respectively.
10 J. Fleck, 'The employment implications of robotics', in *Proceedings of First Conference on Human Factors in Manufacturing*, ed. T. Lupton (IFS Publications, Kempston, 1984).
11 M. Cross and P. Mitchell, *Applying Process Control to Food Processing and its Impact on Maintenance Manpower* (Technical Change Centre, London, 1985).
12 P. Senker, *Towards the Automatic Factory* (IFS Publications, Kempston, 1985).
13 J. Parnaby, 'The design of competitive manufacturing systems', *Proceedings of Fourth International Conference on Systems Engineering*, Coventry Polytechnic, 10–12 September 1985.
14 'A working culture is turned on its head', *Financial Times*, 7 August 1985, p.10.
15 *Industrial Computing*, February 1987.
16 *Financial Times*, 8 August 1989.

17 *Financial Times*, 21 August 1989.
18 National Research Council, *Human Resource Practices for Implementing AMT*, Report of the Committee on Effective Implementation of AMT, National Academy Press, Washington, D.C., 1986.
19 T. Cummings and M. Blumberg, 'Advanced manufacturing technology and work design', in *The Human Side of Advanced Manufacturing Technology*, ed. T. Wall, C. Clegg and N. Kemp (John Wiley, Chichester, 1987).
20 A. Friedman, *Industry and Labour* (Macmillan, London, 1977).
21 See, for example, D. Boddy and D. Buchanan, *Organisations in the Computer Age*, (Gower, Aldershot, 1984), G. Winch (ed.), *Information Technology and Manaufacturing Processes* (Rossendale, London, 1983 and H. Rempp, *Wirtschaftliche und Soziale Auswirkungen des CNC-Werkzeugmaschineneinsatzes* (RKW, Eschborn, 1982).
22 B. J. Jones and P. Scott, 'Working the system', in *Managing Advanced Manufacturing Technology* ed. C. Voss (IFS Publications, Kempston, 1986).
23 S. Prais and K. Wagner, 'Productivity and management: the training of foremen in West Germany and the UK', *National Institute Review*, National Institute of Economic and Social Research, February 1988.
24 P. Senker, cited in A. Campbell and M. Warner, *Management Roles, Skills and Structures*, Research Paper 2/89, University of Cambridge, Engineering Department, 1989.
25 W. Haywood and J. Bessant, *The Swedish Approach to FMS*, Occasional Paper, Centre for Business Research, Brighton Polytechnic, 1987.
26 J. Child, 'Organisation design and AMT', in *The Human Side of Advanced Manufacturing Technology*, ed. T. Wall, C. Clegg and N. Kemp (John Wiley, Chichester, 1987).
27 See, for example, G. Winch, *The Implementation of CAD/CAM*, Warwick Papers in Management, No. 24, Institute for Management Research and Development, University of Warwick, Coventry, 1989.
28 Thompson, *Organisations in Action*.
29 Campbell and Warner, *Management Roles, Skills and Structures*.
30 C. Perrow, 'The organisational context of human factors engineering', *Administrative Science Quarterly*, 24 (1983) pp.570-81.
31 G. Morgan, Images of Organisations (Sage, London, 1986).
32 R. Lamming, 'For better or worse?', in *Managing Advanced Manufacturing Technology*, ed. C. Voss (IFS Publications, Kempston, 1986).
33 Lamming, *Managing Advanced Manufacturing Technology*.
34 S. Smith and D. Tranfield, *Manufacturing Change* (IFS, Kempston, Bedford, 1991, forthcoming).
35 T. Burns and G. Stalker, *The Management of Innovation* (Tavistock, London, 1961).
36 Child, 'Organisation design and AMT'.
37 C. Handy, *Understanding Organisations* (Penguin, London, 1979).
38 C. Argyris and D. Schon, *Organizational Learning* (Addison–Wesley, Reading, Massachusetts, 1970).
39 M. Crozier, *The Bureaucratic Phenomenon* (Tavistock, London, 1964).
40 H. Mintzberg, The Structure of Organization (Prentice–Hall, Englewood Cliffs, New Jersey, 1979).
41 See, for example, R. Hayes, S. Wheelwright and K. Clark, *Dynamic Manufacturing* (Free Press, New York, 1988).
42 O. Sundqvist, *New Technologies in the 1990s – a Socio-economic Strategy* (OECD, Paris, 1988).
43 Cited in J. Ettlie, *Taking Charge of Manufacturing* (Jossey-Bass, San Francisco, 1988).
44 P. Simmonds, private communication.

45   J. Gapper, 'At the end of the honeymoon. . .', *Financial Times*, 10 January 1990.
46   Ettlie, *Taking Charge of Manufacturing*.
47   *Financial Times*, 11 February 1988.
48   'Pressure grows on Ford to conform', *Financial Times*, 11 February 1988.
49   C. Leadbeater, 'ICI seeking formula to alter workers' role', *Financial Times*, 12 January 1989.
50   T. Wall and C. Clegg, 'AMT, work design and performance', *Journal of Applied Psychology* (1990).
51   *Financial Times*, 5 January 1990, p.10.
52   D. Henderson and B. Horsley, 'Commissioning large-scale methanol plants in ICI Agricultural Division', in *Production '78* (Institution of Chemical Engineers, Rugby, 1978).
53   *Financial Times*, 10 January 1990, p.18.
54   E. Medina and V. Cartaya, 'New organizational techniques in Venezuela', *Technovation* Autumn, 1989.
55   E. Miller and A. Rice, *Systems of Organisation* (Tavistock, London, 1967).
56   K. Persson, 'Obstacles, strategies and mechanisms', contribution to Joint Meeting on New Manufacturing Technologies and Industrial Performance, OECD, Paris, 7–11 November 1988.
57   R. Taylor, 'Spoonfeeding a national malaise', *Financial Times*, 30 August 1989.
58   D. Jones, J. Womack and D. Roos, *The Machine that Changed the World* (Ravoson Associates, New York, 1990).
59   R. Taylor, 'Spoonfeeding a national malaise'.

13 Implementation

13.1 DOING IT...

This book has attempted to highlight some of the key challenges associated with moving to a new pattern of manufacturing through the use of new technological opportunities. It has provided some outline blueprints for the successful factory of the future, but one piece of the puzzle is still missing – how to get there. That theme, of implementation, is too extensive to cover here and would properly form the subject of a separate book.[1] However, some brief comments are offered.

Implementing new technology is not a manufacturing version of a shopping expedition in which the main problems are which shops to go to, what products to buy and how to pay for them. Rather, it is a process with a long timescale and many dimensions. Above all, it is essential that any decision about new technology be located within a *strategic* framework.

13.2 Implementing CIM – the need for a strategic approach

Moving towards total integrated manufacturing is not simply a matter of a short-term investment in one or two discrete items of equipment, but rather a long-term philosophy involving technological and organizational components which need to be carefully linked to provide support for the overall business.

All too often, components such as FMS are installed with little idea of the strategic objectives, or of criteria to assess their contribution to meeting these. Where criteria do exist they are often defined in a narrow technical or financial sense, rather than taking in the wider context of the organization's business environment and needs. For example, the success of an FMS might be judged against its cost, its payback and its contribution to cost reduction and output maximization within a small part of production, rather than by increases made to overall organizational responsiveness and agility in a competitive marketplace.

To be effective, a CIM strategy needs to begin with a thorough analysis of the business needs and a clear plan which identifies the basis on which competitiveness will rest in the long and short term. From this the key *order-winning* criteria – flexibility, agility, quality, and so on – can be derived.[2]

The next stage requires a review of existing manufacturing operations, in terms of their local strengths and weaknesses and their fit with this broad strategic framework. Simultaneous with this is the requirement for a thorough exploration of opportunities opened up not only by new manufacturing technologies – such as FMS – but also new or improved manufacturing techniques, such as just-in-time. From this analysis it becomes possible to develop a coherent and appropriate manufacturing strategy for the firm, which will provide the underpinning in order to meet the key criteria in the business strategy.

Within the framework of this manufacturing strategy a long-term CIM/TIM plan can then be developed which identifies the architecture (the layout of different components), the communications between those elements (and the level of sophistication required in such networks), the hardware and software requirements and the underlying organizational infrastructure (including suitable skills, functional support and decision-making arrangements) which will be required. Such a plan will also identify the priority areas and the overall sequence for implementation.

One major requirement in this process is for the development of a parallel organizational development strategy, to ensure that the necessary degree of organizational integration is available to underpin these technical changes. Finally, the strategy can be implemented on a project-by-project basis, moving from 'islands of automation' through to full computer-integrated manufacturing. The advantage of this approach is that it permits the lower-cost and lower-risk features of an incremental philosophy to be retained, but moves the firm forwards within a clear integrating framework.

This approach provides a rationale for making choices and avoids the risk of the installation of expensive and inappropriate systems. The process can be summarized in the following five-stage checklist:

1   Identify business goals and strategies

- What business are we in and why?
- Where are we going to be in five years' time?
- How are we to get there (diversify, acquire, and so on)?

2   Identify relevant product strategies

- What will our products compete on?
- How far do costs have to fall, quality to rise, and so on?

- What will customers be looking for in five years' time?

- What volumes and changes do we expect?

- What level of customization versus standardization do we expect?

- What are the external trends in competitors' products/technologies, and so on?

3   Manufacturing strategies

- What do we need to do to make the products and deliver on the price/non-price factors identified above?

- What implications does this have for processes in use, plant and facilities, human resources, work organization, control and information systems, suppliers, and so on?

4   Define integrated operating strategies

- What strategies do each of the key functions – marketing, design, manufacturing, and so on – need to perform in support of the above?

- How can these be integrated so that they support each other?

5   Define systems and automation strategies

- What systems can deliver on these?

- Automation? JIT? Re-organization? – Which technology (broadly defined) should we select to meet the clearly identified needs?

- Build up to CIM/TIM on a multilevel basis, especially with regard to the information system, which will be both the blood circulating within and between these levels and also the nervous system of the organization as a whole.

13.3   STEP-BY-STEP IMPLEMENTATION

The next task is to begin actual implementation. Here the key lesson is that projects of this scale succeed when there is some form of phased implementation – a step-by-step strategy. Components of this include the use of pilot projects and simulation, and a recognition of the value of the learning process so that time to assimilate and build on such learning is allowed in the overall project plan. For example, consider the case of computer-aided design.

It is possible to see CAD application on the user side as represented by a ladder with several rungs of experience. Initial entry can be made by using CAD to support basic 2-D draughting work. As users become more experienced, more applications programming can be carried out, and on the back of this some work in 3-D using wire-frame systems becomes possible. Further experience and investment will permit complex 3-D work, and the next stage involves linking CAD systems to other production elements. Finally, the possibilities of CIM and electronic data interchange (EDI) can be explored. (Although this model is based primarily on mechanical engineering applications of CAD, it can equally apply in electronics or other sectors where a similar trend towards greater cost and complexity is found.)

The same pattern of development via incremental stages can also be applied to other areas of manufacturing as they converge towards full CIM and beyond. A key benefit of this approach is that it avoids the 'big bang' risks of high cost and complexity – as indicated in figure 13.1.

The same step-by-step philosophy can also be applied to the introduction of just-in-time or total quality programmes; for example, Schonberger suggests a 17-step process moving from simplification and focus on the customer to more advanced applications.[3]

Much of the discussion of technologies throughout this book has highlighted the wide and proliferating choice available to users. This facilitates the implementation of such a step-by-step plan within a longer-term framework, building up from simple, substitution innovation to more highly integrated forms.

13.4 INNOVATION AS A PROBLEM-SOLVING PROCESS

As we have seen, technology emerges in part as a response to particular needs – in other words, as a solution to a problem. So it follows that we can learn from experience in effective problem-solving some lessons which will provide useful guidelines in implementing new technology.

Problem-solving is usually represented as a sequential process involving the stages of recognition, definition, exploration, selection and implementation. It begins with a recognition – an awareness that a problem exists – but this needs to be carefully defined to clarify and focus on the core problem. Without careful attention to the problem definition stage it is easy to engage in solving the wrong problem, or in trying to deal with a symptom rather than a cause. Experience shows here that this iterative process of arriving at the problem definition is best carried out with a variety of information and experience inputs.

In the case of new manufacturing technology there is no shortage of problems to which the technology could be addressed. But without a careful, systematic and strategic analysis or audit of the particular issues

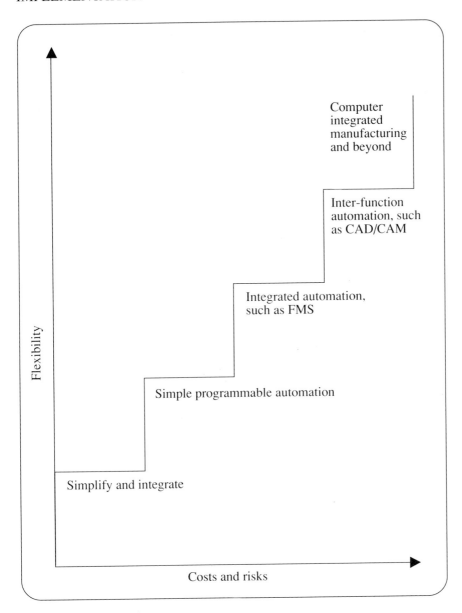

Figure 13.1 Step-by-step progress towards CIM and beyond

involved it is quite likely that the wrong problem will be treated. For example, a typical justification for installing a computer-aided production management system is to try to reduce inventories and improve delivery performance. However, the root cause of these problems is often not related to the need for a faster and more powerful information system to keep track of things more efficiently, but to the presence of an overly complex system that is already in place, trying to keep control over a manufacturing process which has, over time, evolved into something which no one fully understands. The real need here is not for more technology – which would only have the effect of 'computerizing chaos' – but rather for a simplification of the existing arrangement. Once this has been done, the next stage might be to invest in a computer system, but the problem might be amenable to other solutions.

It is also important to take into account the different pieces of information which might help to arrive at a complete and useful statement of the problem to be solved. This may well involve different people, with their own particular mix of perspective, experience and knowledge. For example, in developing a technology strategy, it may be very useful to bring in operators and supervisors from the shop floor, who have day-to-day exposure to the particular key issues which prevent the firm from achieving its competitiveness goals of flexibility, responsiveness, and so on.

Once the problem has been defined, the next stage is to explore the range of potential solutions. This is again an activity which benefits from having a number of different perspectives and, once again, it is an iterative process. Exploring the possible solutions may lead to further redefinition of the problem. Evidence also suggests strongly that the quality of the final solution is improved by spending time generating a range of alternative options here rather than choosing the first and most obvious answer. It may not be – and often is not – the best.

This highlights a typical behaviour pattern in the adoption of new technology. Because of the widespread availability of powerful technology and the perception of major problems facing the firm, the tendency is to look for 'plug in' solutions – the quick technological 'fix'. So, for example, firms requiring greater flexibility will often decide to try and then buy a flexible manufacturing system which, to judge from the label, represents the obvious solution. In doing so they jump straight from a definition of the problem to a solution without adequately exploring alternatives. By contrast, many firms have now discovered that organizational change routes – such as the use of JIT systems – can also solve their problems of flexibility. Rather than buying flexibility, they have explored ways of becoming flexible themselves. Such alternatives are often cheaper and more compatible with the organization, but deciding on them requires considerably more in the way of exploration of the available options.

Once a range of potential solutions have been explored, the next step is to select a particular option. As we have already seen in the case of AMT, making such choices is often a highly uncertain process, particularly in a selection environment where there is limited information or experience, and where there is strong sales pressure from suppliers. Here again the contribution of a range of experience, and information is of value, particularly in ensuring that the selected solution is one which will be appropriate to the broader context in which it will have to work, as well as in solving the problem to which it specifically relates.

Finally the chosen solution has to actually be implemented – introduced into the organization to deal with the problem that was originally identified. One important pay-off from involving a wide range of people in the earlier stages emerges here: the more they have participated in the problem-solving process, the more they will feel a sense of 'ownership' of the solution and a commitment to making it work.

Perhaps the key lesson for the management of technology here is to treat innovation as a systematic problem-solving process and to ensure that sufficient attention is paid to each stage, rather than jumping and taking short cuts in an attempt to speed up the process. The result is likely to be a lower-quality solution.

13.5   IMPLEMENTING ORGANIZATIONAL CHANGE

As we have seen, the theme of organizational adaptation is becoming central to ensuring the successful implementation of advanced technology. In particular, it is clear that the bigger the change the more the present organization will need to shift. The word 'revolution' implies 'the overthrow of the previous order', and so technological revolutions are likely to require significant organizational shifts. Yet the natural tendency of organizations is to resist change, and so success will depend not only on making the necessary adaptation but also on how well these changes are implemented.

In their exploration of the management process in respect of major technological change, Tranfield and Smith argue that there are two types of change, morphostatic and morphogenic.[4] In the former, adaptation is a development along previously established lines, an evolutionary change. This type of approach is suited to firms involved in substitution innovation, where they are essentially making minor improvements to their activities. By contrast, morphogenic change involves a radical shift, a 'quantum leap' in adaptation which involves finding totally new ways of structuring and operating. This type of change is linked to major technological changes such as the move towards high levels of integration.

It is a natural – and valuable – human instinct to resist change, and so any attempt to introduce major shifts of the kind we have been discussing must explicitly take into account this feature of individual and organizational life. Some typical sources of resistance to change in organizations are listed below.[5] Some of these are fairly readily apparent and can be dealt with in a direct fashion; for example, lack of skill can be addressed with relevant training. But others belong in the emotional or behavioural domain and may not be readily visible – and even when they are identified, they may be difficult to deal with:

- **Systemic factors, which operate in the *cognitive* domain**
    - lack of information
    - lack of skills
    - lack of resources
    - lack of support
    - lack of reward
- **Behavioural factors, which operate in the *emotional* domain**
    - anxiety
    - threats to status
    - threats to power
    - fear of failure
    - lack of trust
    - reluctance to experiment
    - culture and norms

Successful implementation depends on managing not only the various aspects of 'traditional' project management, such as procurement, development, construction, commissioning, and so on, but also the process of organizational adaptation. The key theme here is 'parallel implementation' – managing the organization as well as the project. The concerns of such 'process' management can be seen in table 13.1, which sets out a 'force field analysis' for the introduction of technological change. Here the driving forces are those which might be expected to promote the adoption of technological innovation – the kind of factor which is highlighted in investment justification and company strategy statements. These represent the rational basis for change – and it is often assumed that the benefits of making such moves will be apparent to everyone, viewed in an objective fashion and seen to be right for the organization. But the right-hand side lists some of the many resisting forces which can act to retard or even block such change.

It is impossible to provide a prescription for successful management of the organizational change process to support major technological change. As we saw in chapter 12, firms vary enormously in their

Pressures to change	Resistances to change
Market environment	Lack of awareness and understandong
Technology push is catalysing change – firms have to adapt to get best out of their investments	Lack of skills
	Organizational inertia – 'adapt the technology to the existing organization'
	'Sedimented traditions'
	Lack of alternative models
	Supplier pressure to shape organization to fit their technology package
	Politics – defending positions
	Culture – 'the way we do things around here'

Table 13.1   Force-field analysis for organizational change

particular circumstances and so the chances of finding a single 'best' way of approaching the problem are minute. However, it is possible to provide some 'guidelines for good practice' which may facilitate the process, and these include the following:[6]

1   *Establishing a clear strategy at top level* (a process which will in itself involve considerable challenge and conflict in order to get real agreement and commitment to a common set of goals). Once this has been done, the next stage is to communicate this shared vision to the rest of the organization: this will essentially involve a cascade process down through the organization during which opportunities are set up for others to challenge and take 'ownership' of the same shared vision.

2   *Communication.* This is probably the single most effective key to successful implementation, but it requires a major effort if it is to succeed. It must be active, not passive, open (rather than allowing information to flow on a 'need-to-know' basis), timely (in advance of change – the informal communication network will disseminate this information anyway and a slow formal system will undermine credibility) and, above all, two-way in operation. Unless there are channels through which people can express their responses and ideas and voice their concerns then no amount of top-down communication will succeed in generating commitment. Other features in communication which contribute to success are the use of multiple media – videos, presentations, face-to-face

sessions – in addition to traditional memos and noticeboards, and the holding of communication meetings in company time – itself a clear expression of commitment to the project.

3   *Early involvement.* Managers often resist the idea of participation since it appears to add considerably to the time taken to reach a decision or to get something done. But there are two important benefits to allowing participation, and allowing it to take place as early as possible in the change process. First, without it – even if attempts have been made to consult or to inform – people will not develop a sense of 'ownership' of the project or any commitment to it, and may express their lack of involvement later in various forms of resistance. Second, involvement and encouragement of participation can make significant improvements in the overall project design. One of the keys to the success which the Japanese motor and other industries have achieved is the effective mobilization of the creativity of all the staff in the company in solving problems of product and process design. Although this may add somewhat to the time and costs of early stages of the project, improvement and problem-solving here is much cheaper and much more cost-effective than later in the project's life.

4   *Creating an open climate.* It is important that the individual anxieties and concerns can be expressed and the ideas and knowledge held within the organization can be used to positive effect. Once again, this involves generating a sense of 'ownership' of the project and commitment to the shared goals of the whole organization – rather than an 'us and them' climate.

5   *Setting clear targets.* With major change programmes it is especially important to set clear targets for which people can aim. People need feedback about their performance, and the establishment of clear milestones and goals is an important way of providing this. In addition, one of the key features in successful organizational development is to create a climate of continuous improvement in which the achievement of one goal is rewarded but is also accompanied by the setting of the next. In their work on implementation strategies for AMT, Smith and Tranfield identify two discrete phases of target-setting – the 'sprint' and the 'performance ratchet'[7] In the former all the efforts of the organization are focused on achieving a major short-term target – for example, reducing defects by 50 per cent or increasing output by 30 per cent – and everything is directed towards achieving that clear goal. Such a sprint has enormous power in harnessing a widely differing staff to a single goal and brings about the development of a new working culture. But the momentum of

such a sprint cannot be maintained forever, and a second mechanism is needed to ensure that the gains made are consolidated.

5   *Investing in training.* Traditionally training is seen as a necessary evil, a cost which must be borne in order that people will be able to push the correct buttons to work a particular new piece of equipment. Successful organizational change depends on viewing training far more as an *investment*, not only in developing specific skills but also in creating an alternative type of organization, one which understands why changes are happening and is capable of managing some of the behavioural processes involved in change. This requires a substantial increase in the resources devoted to training, extending them to cover broader kinds of input, much of which is devoted to individual development.

Such a prescription is underpinned by a growing weight of research experience. For example, Tranfield and Smith produced a similar model after their detailed two-year research study of organizations implementing a variety of major AMT changes. In the US the National Research Council set up a special committee to explore issues around the effective implementation of AMT.[8] Members of the committee visited a total of 16 sites and concluded that effective implementation depended upon factors such as:

- employment security as a basic platform for change

- a clear business rationale for using AMT and the effective communication and discussion of this at all levels (they particularly stressed 'unprecedented efforts to communicate thoroughly to employees and their representatives the competitive realities of the business, the conditions requiring AMT and the plans for implementing it' – the value of this is not simply confined to gaining acceptance of technical change since it also provides a powerful motivator for major organizational innovation)

- high or equal priority given to planning of human resource issues in comparison with technical and physical issues

- management efforts to effect culture change, and to support and guide the development process

- employee participation in all stages of the implementation process

- broad training that begins before assignment to the project

- openness to learning from one's experience and that of others

- systematic, periodic evaluation

Boddy and Buchanan suggest a similar three-element model for introducing change, based on the '3 Ps', people, process and purpose.[9]

A recent example of such an implementation plan in action was reported by managers from the ICI Pharmaceuticals plant in Macclesfield, UK.[10] For many years the plant had received objectives and a clear statement of its role from the international parent division, but traditional approaches to setting annual objectives failed or produced only short-term improvements. Consequently, in 1987 an attempt was made to change the plant culture to one of continuous improvement. This led to a programme involving and supported by senior management, with elements which included:

- clear basic objectives of improving business performance in relation to priorities set by consideration of the needs of the business and of the people working within it

- a process which was progressively passed down and 'owned' by each distinct level of the management in the organization

- a process which involved everyone on the plant

- a process based on clear understanding, from values and concepts, to strategic statements, to specific achievable action plans, with resources and plans to achieve each objective

- measured improvements in performance as outcomes of the above process

- a different style of management which focused on respect for people and provided them with the opportunity to develop in their jobs

13.6   SIMULTANEOUS CHANGE

A theme which emerges strongly in research studies of successful implementation of AMT is the need for simultaneous organizational and technological change. This is highlighted in many studies of particular technologies; for example, Adler and Helleloid,[11] and Winch and Voss.[12] Unlike earlier models in which it is often suggested that organizations adapt to technology in a sequential process, success with AMT appears to require a model of parallel change. Leonard-Barton,[13] for example, speaks of the need for a process of 'mutual adaptation', while Ettlie[14] argues for 'synchronous innovation'. His prescription to firms is to 'stay on the diagonal', commenting on the need to retain the balance between organizational and technical innovation, whether dealing with small or large increments of change. A generalized model for such change is offered in figure 13.2.

Some experience with firms which have adopted a high-technology strategy to deal with problems, but which have failed to take adequate account of the organizational shifts involved, supports this view. By the same token, firms which have followed the Japanese approach, stressing organizational change as a prerequisite for technological change, have tended to be more successful with their subsequent experience, but are in danger of lagging behind if they fail to invest in AMT as the next stage of this process.

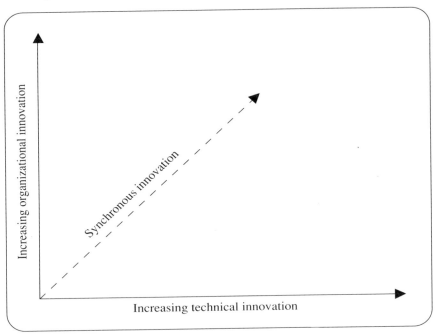

Figure 13.2   Simultaneous innovation

This discussion lends strong support to the views advanced by Fleck[15] and Tidd[16] who suggest that current generations of AMT represent 'configurational technologies' which are evolving as part of an interactive process involving organizational and technical dimensions, rather than as part of a predetermined technological trajectory.

13.7 THE LEARNING ORGANISATION

A final point, concerning the process of change and also the issue of maintaining long-term momentum, is worth discussing here. In much of the above reference has been made, explicitly or implicitly, to the need for organizations and individuals to learn, to be prepared to make mistakes and to use the information so gained to improve future performance – often by making major changes in the way in which they approach issues of organizational structure and process.

One way of looking at organizations is as living systems, interacting with their environment and trying to survive and develop. Such a model implies the need constantly to adapt or die, and this provides an analogy with the challenge facing manufacturing organizations at present. Clearly, any failure to appreciate the emerging realities of the highly competitive and turbulent environment may lead to a loss of competitiveness and possible death of the enterprise – as other leaner, fitter and more agile competitors adapt better.

But the challenge is not to achieve a single adaptation followed by stagnation. Instead, there is a need to build in a system of continuous renewal, of constant responsiveness to the environment coupled with sufficient internal stability to maintain the organization. So any technological solution should not be seen as a short-term fix but as part of a process of continuous improvement – and no matter how good it may be, there is still room for further improvement.

This once again highlights the point about the total organization. If – as many now argue – we are moving from an era of capital intensity to one in which knowledge and problem-solving ability becomes as important a feature as fixed capital equipment, then the ability to become a learning organization, constantly developing and improving this knowledge resource, becomes critical. But – despite developments in the field of artificial intelligence – such creativity and tacit knowledge cannot be embodied in machines to a sufficient level. Significantly, this marks a reversal of the traditional labour process which culminated in Fordism and which aimed to marginalize and eliminate people from the factory, replacing them by machinery. In the learning organization it is the total system – especially the human resources at all levels – which will contribute the competitive advantage in the future.

Notes

1   Good discussion of the issues of implementation can be found in E. Rhodes and D. Wield, *Implementing New Technologies*, (Basil Blackwell, Oxford, 1985), D. Boddy and D. Buchanan, *Managing New Technology*, (Basil Blackwell, 1985) Oxford,, J. Ettlie, *Taking Charge of Manufacturing* (Jossey-Bass, San Francisco, 1988), and S. Smith and D. Tranfield, *Manufacturing Change – Guidelines for Managers* (IFS, Kempston, Bedford, 1991, forthcoming).
2   T. Hill, *Manufacturing Strategy*, (Macmillan/Open University, London, 1985).
3   R. Schonberger, *World Class Manufacturing*, (Free Press, New York, 1986).
4   Smith and Tranfield, *Manufacturing Change*.
5   This table is based upon a version described in R. Plant, *Managing Change and Making it Stick* (Fontana, London, 1987).
6   These guidelines are based upon a model developed by Smith and Tranfield and described in their book, *Manufacturing Change*.
7   Smith and Tranfield, *Manufacturing Change*.
8   National Research Council, *Human Resource Practices for Implementing AMT*, Report of the Committee on Effective Implementation of AMT, National Academy Press, Washington, D.C., 1986.
9   D. Boddy and D. Buchanan, *The Technical Change Audit*, (Manpower Services Commission (now the Training Agency), Sheffield, 1987).
10   Institution of Chemical Engineers, *Managing Change in the Process Industries*, I. Chem. E. Symposium Series, (Rugby, 1989).

11 P. Adler and D. Helleloid, 'Effective implementation of integrated CAD/CAM', *IEEE Transactions on Engineering Management*, EM34(2) (1987), pp.101–7.

12 G. Winch and C. Voss, paper presented to British Academy of Management, September 1989.

13 D. Leonard-Barton, 'Implementation as mutual adaptation of technology and organisation', *Research Policy*, 17 (1988), pp.251–67.

14 Ettlie, *Taking Charge of Manufacturing.*

15 J. Fleck, *Innofusion or Diffusation?*, Working Paper, Edinburgh University.

16 J. Tidd, *Flexible Manufacturing and International Competitiveness* (Frances Pinter, London, 1990).

14 Common Threads

14.1 BEST-PRACTICE FACTORIES

One of the terms increasingly used in discussing manufacturing performance is 'world class', describing companies which have reorganized and invested in technological and organizational develop-ment to a point at which they are able to compete with any firm in the world.[1] Analysis of such firms reveals a striking similarity, suggesting that, while firm size and sector still exert an important influence, a common blueprint for success emerging.

For example, consider the following cases drawn from a UK vote for 'Britain's best factories' in 1988 and 1989[2].

Yamazaki Machine Tools, Worcester

This new plant cost around £35 million and deploys FMS, MRP2 and a variety of other self-developed systems. Quality performance is higher than that achieved in the Japanese parent's plants and inventory levels are only 20 per cent of industry average. The company has tried to compete on high quality and low cost, as does every other machine tool builder. But, in addition, Yamazaki are trying to demolish the distinctions between broad range and niche manufacturing; that is, increasing variety and reducing lead times. This involves them in a major, long-term strategy for design, manufacture and marketing; for example, whereas most European firms offer between 10 and 15 models, they offer 55 on lead times of two months, compared to industry standards of six months.

High variety is supported by short lead times and this is achieved in turn by being responsive – through flexibility in machinery and people. At present the plant employs 180 people making 1000 machine tools a year, with a turnover per employee of nearly £400 000. Within the plant there are no unskilled jobs because low-skill jobs have been automated, while complex manual assembly tasks as handled by flexible, multiskilled staff. This arrangement is strongly supported by training and all staff are salaried, pensioned and health insured. In short, labour is seen as an asset.

Within the plant there is a strong problem-solving emphasis and a clear continuous improvement culture.

Linn Products, Eaglesham near Glasgow

This small manufacturer of hi-fi systems was an early user of all major computer-aided technologies and can boast that it has more terminals than people. It succeeds in competing with Japanese manufacturers on an entire portfolio of hi-fi systems, not just on a narrow product range. Unlike many domestic firms, it has a high commitment to the future of its products, with 15 per cent of turnover spent on R & D. The manufacturing facility is highly automated but within a clear strategy.

Production was originally carried out in lines but the firm realized their limitations and shifted to the idea of cellular manufacturing, using a single-station build concept, with one person responsible for assembly, test, packing and signing off. This approach (which is backed up by extensive training) builds in quality and involvement, essentially recreating elements of the old craftsman model. In structural terms, there is no hierarchy, and management are seen as essentially a support function.

The plant uses very high technology – AGVs, CNC, close coupled logistics and a flexible layout configured around product mix. Another key element is that everyone in factory is 'coupled' to the outside world; for example, customer visits to and from plant are a regular feature. Emphasis is placed on designing quality into product with a strong in-house design team.

Not surprisingly, this pattern of organizational and technological change has enabled them to achieve significant benefits, including a WIP inventory level of almost zero and lead times down from 7–8 months to 1.5.

Lucas Diesel Systems, Sudbury

This plant, producing fuel injection equipment for diesel engines, is part of the Lucas Group. In 1984 the company experienced a major crisis of competitiveness, with falling market share, poor quality and reliability, declining customer relations, and so on. As a result of an analysis of that situation, a major strategic change programme was embarked upon which involved extensive organizational development but relatively little investment – only £1.5 million over four years.

Essentially, this programme involved the reorganization of the plant into separate manufacturing cells. Production was simplified by creating three minifactories, each focused on a small coherent family of products. At the same time the management hierarchy was reduced from seven layers to three. On the employee side, jobs were transformed from those requiring 'human robots' to those needing problem-solvers with flexibility within and across cells. The necessary skills development was backed up by training and by placing stress on common ownership of manufacturing problems. The culture moved from a traditional crisis management factory, oriented to output, to one which stressed continuous improvement.

The results include massive cuts in lead times (300 per cent), improvements in delivery reliability, increased inventory turns (from 22 to 34 times and rising), higher quality (up by 12–15 per cent), savings in space through better layout and a productivity improvement of around 50 per cent. Perhaps the most significant achievement here has been that almost all of this benefit came without extensive capital investment, but with systematic challenge to the existing organization and management of production.

JCB, Rocester, Staffordshire

This company produces mechanical excavation and related equipment and has developed a manufacturing strategy based upon competing on responsiveness and quality. (Their definition of quality is wide, covering not only quality of product built but also its timely delivery, its design, its cost and the level of service which is offered to support it.) This strategy is supported by investments in JIT, TQM plus appropriate automation – CAD/CAM, robots and CNC – but not technology for its own sake. The culture is characterized as one of continuous improvement and constant change, although not all of the increments are large. Many benefits have come, for example, from fitting castors to machine tools to enable them to be rearranged easily. As the factory manager commented, 'I believe a great deal of harm has been done by people pursuing the Holy Grail of CIM.'

The benefits are again clear. The company has the highest figures for turnover per employee, profits and investment in the industry. Overall turnover increased by 300 per cent and labour productivity rose by 125 per cent over a ten-year period. Inventory management is also better, with a stock turn-up from 3.2 to 15.3, and there is a higher quality supply base, reduced from 730 to less than 400. Manufacturing lead time, from initial order to final delivery, is now down to eight weeks.

Black and Decker, Spennymoor

This manufacturer of power tools has achieved levels of manufacturing performance at least as impressive as its US parent, which has been the subject of much discussion. Two concepts dominate – total quality management and total customer service – and these are implemented through bringing people and automation together under this philosophy. In beginning the process of change, they did not cut their product range or go for robots but redesigned products for families – systems such as transmissions, engines, and so on – and 'modularized' them. They then applied JIT plus appropriate automation, and gradually built up to their current flexible assembly system, which is supported by a sophisticated computer-based information system.

This approach now enables them to offer a high variety in their product range range – 29 different types of drill, for example. Over eight years lead time has been cut from five to two weeks, inventory turn has increased from 8 to 29, scrap has been reduced by 400 per cent, assembly rejects have been reduced by 370 per cent, the cost of quality has halved, while output and sales per employee have both increased, by 400 per cent and 440 per cent respectively.

Sony, Bridgend

This manufacturer of television sets is one of the longest-established Japanese-owned plants in the UK. It attributes much of the improvements to extensive application of Japanese manufacturing techniques such as JIT, total quality management and the use of multiskilling and continuous improvement, involving all the workforce in problem-solving. In addition, the company has made extensive investments in MRP2, EDI and flexible manufacturing systems, supporting such technology with extensive training. Significantly, all workers are also salaried and there are fewer hierarchical levels in the plant than in many European competitors.

Benefits from this approach include increased output, much of it exported (75 per cent of the 690 000 sets in 1988), reduced costs, 99 per cent delivery performance, a 7000 per cent fault reduction in four years, low inventories (WIP is around 2.83, finished goods at 0.65 days, shopfloor inventory 0.6 days), and increased productivity (with output per worker up by 450 per cent).

ICL, Ashton-under-Lyne

This computer manufacturer makes extensive use of CIM technologies, including two linked FMS facilities capable of handling any products within size envelopes of 0.5 and 1 cubic metre respectively, MRP2, CAD and company-wide application of JIT and TQM. Their manufacturing strategy places emphasis on short lead times, high quality and service – all at low cost.

In order to achieve this they worked hard to design labour out of their products, including activities in areas of assembly and test, but their 'manufacture as service' concept also requires a high level of involvement and problem-solving support. Consequently, they have invested in a very high level of training (around eight days minimum per employee per year) and skills development, and have changed the payment system so that every employee enjoys staff status and progression is based on an annual review of performance.

Benefits include reduction in lead times from three to ten weeks (for batches of one if necessary), increased inventory turns (3.2 in 1985 to 12.7 in 1989), better delivery performance (96 per cent, now on-time) and annual cost reductions of 10 per cent.

Toshiba, Plymouth

Toshiba is another television set manufacturer which has transferred many Japanese practices, including JIT and TQM. Emphasis has been placed on three main areas of change – supplier relations, product and process rationalization and human resources. On the organizational side they have reduced the number of levels in the hierarchy in the plant to five, and labour now works on a team approach. Supervision has given way to team leaders and the range of tasks carried out by these teams has been extended to include extensive problem-solving activity. All employees are covered by the same salary scheme and there is considerable time flexibility, with a seasonal working pattern involving shorter working weeks in the low season – an approach which has helped to cut labour turnover.

This organizational change is paralleled by extensive use of technology, based around MRP2 for materials management, and around flexible assembly systems for actual production.

Benefits include a 100 per cent delivery record, reduced inventories (WIP has been cut to 3.27 days), no goods inwards inspection and greater flexibility, with batch sizes cut to 500. Of particular significance is that, on an output of nearly half a million sets per year, they have managed to double their share of the market, much of this due to quality improvements.

Rank Xerox, Mitcheldean, Gloucestershire

This plant, building photocopiers, employs relatively low technology in actual manufacture but extensive use of IT in coordination and inter-firm dealings, based on MRP2 and EDI. In particular, EDI has helped them to reduce inventory to 25 days (still falling) and to concentrate on four key supply partners. Significantly, they threw out an earlier generation of computer-based control because it limited their flexibility.

Within the manufacturing process emphasis is placed on people as a resource, with extensive use of JIT and TQM, embodied in a multiskilled cellular manufacturing approach. This involves two operators per station (a multiskilled assembler and a quality tester) and has the advantage of providing instant quality feedback.

Benefits include a 99 per cent delivery record, a better market reputation on quality, inventory down to 25 days, 35 per cent zero defect weeks, 50 per cent space saving, improved forecasting and a 20 per cent reduced new product development cycle through teamwork.

Oxford Automotive, Oxford

This small automotive components specialist operates with a cell-based approach, using JIT and TQM principles and multiskilled workers. Coordination is via MRP2 and inter-firm dealings make extensive use of EDI.

Benefits include a 100 per cent delivery record, lead times cut from six weeks to four days, inventory reduced to 20 days (WIP to 3.5 days), space savings of 50 per cent and a 400 per cent increase in profits. Much of this was due to the effective turnaround of reputation in the marketplace, breaking away from a former poor quality image to one of reliability and service.

14.2   KEY FEATURES

There are a number of key features common to these successful manufacturing organizations in the 1990s, and to other success stories:

- a clear strategic framework for the business

- manufacturing strategy a priority issue

- strong customer orientation throughout the business

- emphasis on total quality throughout the organization

- extensive investment in training and development at all levels

- minimal hierarchy and extensive decentralization and local autonomy

- continuous improvement – learning organization

- appropriate use of automation technologies

- staff status for all, and reduction or elimination of different gradings

- partnership with suppliers

- employee involvement and problem-sharing

14.3   A NEW BLUEPRINT?

What these and many other similar stories indicate is that significant change in performance at the level of overall business effectiveness can be achieved. For all of these firms, the innovations have contributed not just to improving local efficiency of a particular operation or area but to the operation of the *whole* system. Although there is some similarity in the tools and techniques used, there is no single prescription for such success. Nor is it confined to a particular sector or size of firm; the spread of experience suggests that moving to such new models is an option open to any firm prepared to take a strategic approach to changing the way it operates.

This also raises another important point regarding Japanese success as a manufacturing nation in the second half of this century. Many of the firms in the above list are Japanese-owned, although operating in the UK and with a high proportion of UK personnel. Whatever arguments can be advanced for a cultural base for Japanese manufacturing success need to be set against the fact that Japanese-style approaches can be made to work successfully outside Japan. (Schonberger makes a similar point regarding Japanese-owned plants in the US.[3])

Such transferability suggests that the key to success is in the adoption of alternative models of factory organization and management. These may have been adopted earlier and become more highly developed in Japan, but they are, fundamentally, international in nature and open to variation and experiment. Just as the mass production models of Henry Ford began life in Detroit but eventually permeated the entire manufacturing world, so the new models which were applied first in Japan are become more and more widely distributed and varied.

We began this book by looking at the car industry, and it is useful to examine what has been happening in that industry. As the results of the International Motor Vehicle Programme study clearly show, the basis for competitiveness has moved on a long way from the days of Henry Ford and Alfred Sloan. The emerging model of best manufacturing practice is one which they term 'lean manufacturing'; so-called because it requires much less input to produce a high-quality product than in traditional factories – less labour, less material, less time, fewer problems, and so on. Lean manufacturing undoubtedly had many of its roots in the experience of the Japanese car industry in the post-war period, especially with the evolution of just-in-time and total quality management ideas. But, just as the Ford system took much longer than Ford's own lifetime, and spread widely beyond its mid-Western US origins before it reached its full potential, so the lean manufacturing system is still evolving and developing – and there are opportunities for its application throughout the industrialized and the developing world.[4]

As suggested at the start of this book, it can be argued that we are moving into a new techno-economic paradigm as we approach the twenty-first century, in which new rules governing best practice in manufacturing are emerging. The actual specification or blueprint for the factory of the future is still hazy, but we do know that the old models based on the Fordist paradigm are becoming increasingly inappropriate. What the above cases indicate is that the process of moving to a post-Fordist paradigm has begun, and that we already have some clues about the shape of the twenty-first century factory. The prime task now is to continue the process of experimentation: we know what no longer works but we still have to discover what sorts of effective alternative might be viable.

To do this requires, above all, a commitment on the part of manu-facturing organizations to continuous change and development. In short, they must become learning organizations, willing to try new things out, to learn from mistakes and to take risks. At the heart of this process they need a mechanism for capturing and retaining knowledge derived from this learning, for it is increasingly going to be knowledge rather than physical technology in products or processes which confers strategic competitive advantage. And this requires major reconsideration of the role of people in manufacturing organizations. Whereas in the old paradigm people were essentially an adjunct to technological systems, a necessary expense which future developments tried to marginalize and distance, the new one is increasingly dependent upon human intervention to provide flexibility, creativity and the ability to capture, retain and pass on knowledge.

A crude but powerful metaphor for the Fordist factory is one which sees people essentially as cogs in a machine – an image which Charlie Chaplin immortalized in his film 'Modern Times'. By contrast, our metaphor for the post-Fordist paradigm must be one which puts them, by virtue of their natural flexibility and creativity, at the brain of the firm, the key to the learning organization of the future.

Notes

1 For example, see R. Schonberger's 'honor roll' in his book, *World Class Manufacturing* (Free Press, New York, 1986).
2 'Britain's best factories', *Management Today*, September 1988 and November 1989.
3 Schonberger, *World Class Manufacturing*.
4 For a full discussion of this, see J. Womack, D. Jones and D. Roos, *The Machine that Changed the World* (Rawson Associates, New York, 1990).

INDEX